T0179606

# Superior Products

"Seventy Five Years of
Product Innovation Experience"

GORDON BRUNNER & WILLIAM JAMES

SUPERIOR PRODUCTS

©2023 GORDON F. BRUNNER

ISBN: 979-8-35092-341-4

# GORDON BRUNNER

Gordon Brunner is a well known and admired R&D leader, having spent a 40 year career at Procter & Gamble, and credited with accomplishing the launching of dozens of new and improved superior products, as well as major advances in systems, culture, reward systems, and leadership principles for a global R&D organization.

He was the Chief R&D Officer for 13 years, worked for four CEO's, and was the first and only R&D employee to be appointed to the Company's Board of Directors, where he served for 9 years.

Gordon Brunner was an avid golf caddie which led to an Evans Scholarship at the University of Wisconsin. He graduated in biochemical engineering, and later obtained an MBA from Xavier University.

During his career, he was involved with five new-to-the-world multi-billion dollar brands, all behind major technology achievements. This included Liquid Tide & Ariel, Pantene, Febreze, Swiffer, and Actonel for osteoporosis. There were some fifteen other multi-hundred million dollar successful innovations and superior brands. P&G's outstanding technical accomplishments were recognized in being awarded the US Medal of Technology in 1995, and where Brunner accepted the award for the Company.

He was a student of R&D organization culture, structure, and effectiveness, and instituted major changes which have become global benchmarks. This included researcher reward systems and career path structure, "open" innovation, global internet innovation communication, and an internal "new venture" project structure. His achievements and perspectives were broadly recognized, and among many awards, he was named the YEAR 2000 MEDALIST by the Industrial Research Institute.

His perspective and experience were applied in a variety of large and small company boards, external new venture companies, academic research commercialization, and entrepreneurial product launches.

He lost his dear wife after 44 years of marriage, but is blessed over the last eleven years with a fan-tastic & loving partner, Mary Ellen, and the joy of three daughters, and eight grandchildren.

# WILLIAM (BILL) JAMES

Bill James is a chemical engineering graduate of Case Western Reserve University and the University of Cincinnati. He started his P&G career in Food Product Development reporting to Gordon Brunner. He spent 36 years at P&G with 20 years in the research labs rising to Director of Food Product Development. In 1985 he became the first Director of Worldwide R&D Human Resources reporting to CTO Wahib Zaki and subsequently to Gordon when he succeeded Zaki. He later became the Global R&D Chief of Staff continuing to report to Gordon Brunner.

Bill has had a life-long interest in understanding the key factors that constitute superior product development. As Gordon Brunner's Chief of Staff, he worked with Gordon on the design and implementation of the many new concepts that were implemented into the R&D structure. This, of course, included the extensive communication task necessary to sell the projects within the Company and with the global R&D organization. He has been heavily involved in R&D professional groups, giving lectures and authoring articles in the Industrial Research Institute's Research & Technology Management magazine, and the Center for Innovation Management's journals.

Within his responsibilities as Head of R&D Global Human Resources, Bill led the establishment of the Victor Mills Society, an industry acclaimed top technologist professional society, honoring the top innovators in the Company. He then led the effort to create a sin-gle-track career path for all 7,000 technical members of the Company's worldwide R&D staff. It recently celebrated 30 years of success-ful experience.

When Bill isn't absorbed in innovation studies he and his very tolerant wife, Mary Jane, enjoy both domestic and international travels. Trips to Seattle and Santa Fe provide opportunities to stay in touch with his adult children, in addition to his daughter and grandchildren in home town Cincinnati. Bill is an experienced furniture builder and has constructed the storage and display cases for his wine collection, along with many other furniture pieces. Relaxation includes a good spy novel, a favorite Bourbon and playing along with Coltrane, Brubeck or Mingus on his alto sax.

# JOHN E. PEPPER
# A PERSONAL REFLECTION

## CEO and/or CHAIRMAN, PROCTER & GAMBLE
## (1995-2002)

Gordon Brunner and Bill James offer future generations of leaders a singular gift in this comprehensive and extremely well researched set of case studies and principles documenting the preeminent importance of outright consumer product superiority and how to achieve it.

Their book derives its strength from the granular, real life learnings from two men, Brunner and James, who spent a combined 75 years leading product innovation at Procter & Gamble. They reveal how they created environments and systems and a culture fostering the development of breakthrough technologies which have created and sustained global leadership brands like Tide, Pampers, Pringles, Pantene, Always, Actonel, Bounty and many more. The stories they share on how these brands achieved success have in the main never been told before. They carry vital lessons for the future.

This is not an all "good news" story. Brunner and James critically examine failures which occurred under their watch and they mine the lessons for the future to be drawn from them.

A particular light this book sheds is the singular importance of strong personal leadership starting at the top and including all line managers and R&D management. This leadership is driven and inspired by the "do-or-die" conviction in the importance of achieving irresistible performance superiority.

Everyone pays lip service to the importance of product superiority. But words won't cut it. Product superiority must be demanded. It must be rewarded. It must be celebrated. And organization structures and processes need to be imagined and implemented to achieve it. This has to be led by senior line and R&D leadership, right up to and including the CEO and Board of Directors.

Brunner and James emphasize through their stories the irreplaceable role of commitment, conviction and courage and the willingness to back a big idea.

Put simply, their book recognizes that the reason P&G and other similar consumer goods companies exist is to provide consumers with brands which meet their needs better than anyone has before—and doing so better tomorrow than today.

To be clear, Brunner and James emphasize that sustaining corporate and brand leadership over the years, thereby providing excellent returns to shareholders, demands superior performance from all the company's functions—Marketing, Manufacturing, Finance, Product Supply, Engineering, HR—all working hand in hand against a common purpose and set of strategies and goals.

However, they make the compelling and convincing case that superior product performance as experienced by the consumer is the absolute foundational requirement. This will come as no surprise to any student of P&G's history. The future always has been and always will be created by new technologies and performance benefits that redefine a category and often create an entirely new one.

# PREFACE

In our book, we are putting forth our learnings from a lifelong interest in innovation, and careers in product development with the world's largest and oldest consumer product company - Procter & Gamble. We provide our view of the ideal innovative work environment, as well as the critical aspects of the innovative process.

We tell the stories behind the development of some of P&G's most iconic consumer products - stories that for the most part have not been previously told. One of us, or both in some cases, were involved in some way with each of the stories. We attempted to include key events, and key people involved in the innovation. In recent years, we actively interacted with many others involved in the projects to gain additional perspective. **However, we know we were not able to capture or document every significant event or outstanding individual contribution**.

Each of the product innovations had a full team of critically-skilled people behind them to make the innovations happen. We applaud the contributions, and wish the scope of the book could have acknowledged all their fine individual work. But, we hope that this documenting of these major innovations will bring appropriate pride and positive memories to all involved in the achievements

Essentially all of the projects have beginnings 20-60 years ago, so memories may have missed some details......but not the fundamental story behind these product innovations.

It's our intent to focus this book and individual stories on the PRODUCT. There is no doubt that every successful product had to be supported by outstanding work from all the other company functions: MARKETING, MANUFACTURING & PRODUCT SUPPLY, SALES, LEGAL, GENERAL MANAGEMENT, as well as our external partners such as advertising agencies and suppliers. Every one of the product innovations had world class contributions from these ESSENTIAL colleagues.

*Gordon Brunner and William James*

# FORWARD

## JON R. MOELLER
### PRESIDENT, CEO, & CHAIRMAN, THE PROCTER & GAMBLE COMPANY

P&G serves consumers with a portfolio of trusted and consumer-preferred brands that improve consumers' lives around the world. This has been true throughout the 186 years of our history, and it will continue to be true behind our success for decades to come. Our brands deliver Irresistible Superiority across product, packaging, brand communication, retail execution, and consumer and customer value.

We focus on categories where product performance is a significant driver of brand choice. And there is no doubt that the significant product innovations in our Company's history enabled the creation of multibillion-dollar brands and continue to spur our progress today. Further, I am convinced our brands need continued product innovation to consistently deliver the highest level of performance, maintain their competitive edge, and develop the categories in which we compete.

An in-depth understanding of the past is a powerful tool that we must draw on to shape programs to continue to grow in the future. Gordon Brunner and Bill James had extensive career experiences in product innovation at P&G. They've captured very valuable learnings on the product innovation process and unique and first-time historical documentation behind the development of many of our great Company products. These learnings and experiences have important relevance to our innovation approaches today.

# Contents

## SECTION FIVE
## "COMPANY STRUCTURES FOR MAJOR INNOVATION"

# SECTION SIX
## "FOSTERING CONNECTIONS FOR INNOVATION"

# SECTION SEVEN
## "THE BOTTOM LINE"

# APPENDIX
## P&G CEO TALKS

# INTRODUCTION

1. WHY DID WE CHOSE TO WRITE THIS BOOK?

2. THE PROCTER & GAMBLE COMPANY AND OUR ENTRY

# 1.

# WHY DID WE CHOOSE TO WRITE THIS BOOK?

## "SEVENTY FIVE YEARS OF UNIQUE PRODUCT DEVELOPMENT AND INNOVATION EXPERIENCES!"

 From the time I graduated from the University of Wisconsin, Biochemical Engineering degree in hand, to my current role as an active Kibitzer to corporate R&D leaders, including the current P&G CTO, I've been involved in product innovation. In retirement I've continued as the chairman of the Equipment Standards Committee of the PGA Tour, wrestling with golf equipment's effect on the game, particularly driving distance. I've spent the last 20 years as chairman of the Wisconsin Alumni Research Foundation commercialization committee, dealing with getting to the market those great technologies emerging from the University researchers. I've been on big and small Company boards, and have been active in private equity, and in actively helping several entrepreneurs launch their products. In retrospect I've been developing products for the market all my professional career. It must be in my DNA.

I was a 40-year employee at Procter & Gamble. I was Chief Technology Officer for 13 years, and a member of the Board of Directors for 9 of them. I worked for 4 different CEO'S - all TITANS OF AMERICAN INDUSTRY - John Smale, Ed Artzt, John Pepper, and Durk Jager. Their approach to managing Research & Development varied quite a bit. But their commitment was unwavering. I learned a lot working with each of them and appreciated the support I had from all of them.

P&G had a very diverse line-up of product categories: laundry, paper, beauty care, foods, beverages, health care and pharmaceuticals During my tenure, there were experiences with more than 10 different product categories and I tackled the challenges of creating, improving and scuttling products in all of them.

We established superior product performance and consumer acceptance goals for each of our products. We drove major advances in each of our existing brands, and targeted the development of major new brands.

During my career, I was involved with five new-to-the world, multi-billion dollar brands…LIQUID TIDE, PANTENE, FEBREZE, SWIFFER, and ACTONEL for osteoporosis. Each product involved the development of new technology, and several involved the utilization of existing technology from another category of business. The development of each involved highly unique learning.

Beyond these five brands, I was involved with some ten new and successful smaller multi-million dollar ones. All sported significant product advantages and unique delivery approaches. This included BOUNCE dryer-added fabric softener, PERT PLUS shampoo, CREST WHITE-STRIPS, THERMACARE heat pad pain treatment, GLAD stretch and odor-controlling garbage bags, GLAD PRESS & SEAL wrap and others.

P&G experienced their share of disappointments: SOFT-BATCH COOKIES, CITRUS HILL ORANGE JUIICE plus the failure of OLEAN, the first zero calorie food fat .

As time went on, I began to realize more and more that I had been blessed with a "one-of-a-kind" career in consumer product innovation and needed to capture it. Further, although there is much appropriately captured and written on Procter & Gamble's management, business results, and marketing of its brands, **there is very little written about the innovation structure in the Company, or about the critical engine behind P&G's success - innovative and consumer-preferred products. That's what this book intends to do.**

P&G has a powerhouse of a technical organization. During my tenure, we had a fully global Research & Development organization with 22 different technical centers, 7,000 total researchers and a budget of almost $2 billion dollars.

In Europe we were the first American company to move from local country products to Euro-products, and I spent seven years in that very important and successful effort.

We established one of the the first American technical centers in Japan..and also in China. We evolved to having 40% of Company researchers outside the U.S.

We were at the leading edge in moving the Company from an American to a global Company, establishing global brands and capturing the learning from our unique approach.

We created a very successful internal NEW VENTURE program, and also created an internal EMPLOYEE RECOGNITION structure for rewarding and motivating researchers for innovation achievements.

We moved to an "OPEN INNOVATION" strategy which involved the complete opening of our patent portfolio. We created partnerships with suppliers and other external companies to conduct innovation together, urging our researchers to embrace "Connect & Develop" rather than just "Research & Develop".

We went through the challenges of investment levels, the need for Corporate R&D, the involvement of the CEO & the Board of Directors, succession planning and the important learning from each of these experiences.

NET....I think this unique career experience provides major learning that has lasting value. If we didn't attempt to capture and share this. it would be lost forever. There is no question that achieving and sustaining superior and Big Edge PRODUCT INNOVATIONS is critical not only to P&G...but corporations broadly. But it's extremely difficult to achieve. I had experiences and challenges most product innovation leaders have not even thought about. Most often, descriptions of the innovation process are explained with charts, complex diagrams and academic verbiage. My approach will be straight forward principles and real-life examples, experiences and results.

I'm very fortunate to have a long-time colleague, Bill James, join me in reflecting on product innovation and compiling this book. Bill is a chemical engineering graduate from Case Western Reserve University, and started at Procter & Gamble in Food Product Development a couple years after I

did. He subsequently spent 36 years at the Company and worked with me in the 1960's on food development projects, and later in the 1980's and 90's when I was CTO, as Global R&D Chief of Staff.

During his career, he was a director of Food & Beverage Research & Development and led the work over the family of P&G coffee, orange juice, prepared mix, shortening and oil, Olestra, peanut butter and Pringles brands. He has a deep interest in product innovation, has been a leader in R&D professional groups, and has authored articles for the Industrial Research Institute, Research-Technology Management magazine, and Center for Innovation Management at North Carolina State University.

As Chief of Staff, Bill also had responsibility for global R&D Human Resources. Bill led the formation of the Victor Mills Society, an industry acclaimed top technologist Society award, as well as a widely adopted Technical Career System - a technologist career path structure for industry, which has survived through today, unlike countless past failed industry dual-ladder attempts.

Like myself, from his career experiences in product innovation, he also has stories and learnings which need to be told.

**WE HOPE THIS BOOK PROVIDES HIGHLY INTERESTING LEARNING, AND STIMULATES THE UTILIZATION OF THE LEARNING AND EXPERIENCES WE SHARE…TO FOSTER and SUSTAIN CORPORATE PRODUCT INNOVATION,**

# 2.

# THE PROCTER & GAMBLE COMPANY AND MY ENTRY

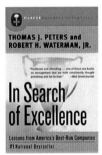

Robert H. Waterman in his best-selling classic, "IN SEARCH OF EXCELLENCE" - a deep study of 43 of the US's best companies, reflected:

*"Until I started interviewing in Cincinnati, I thought Procter & Gamble as a fine consumer-products company - I still do.*

## But now, first and foremost, I think of P&G as a technology powerhouse

*Of course its people are great marketers. But for them, marketing would be an empty discipline if they didn't have something **outstanding** to sell. When they go to market with a new product.......it's far better than anything else out there"*

· · · · · · · · ·

PROCTER & GAMBLE is the largest consumer goods Company in the world. It was started in 1837. Today it has maintained its leadership status for 186 years. PRETTY REMARKABLE.!

The company was founded by William Procter, an English candle maker, and James Gamble, an Irish soap maker in Cincinnati, Ohio. From the start, their focus was on superior quality products. During the Civil War the candles they supplied to the Northern troops were of such superior quality that the river boat shippers put a star on the wooden cases to distinguish them.

Forty two years later, the first product "innovation" was launched - IVORY bar soap. It was a purer ( 99 44/100%) soap, starkly white, milder to the skin, floated in water, and was premium priced. Company lore has it that IVORY actually was a bit of serendipity in that air was accidentally injected into a very pure soap mixture, creating a very mild soap bar that floated, and it became the Company's first unique leading brand. Looking deeper into the archive one can find evidence that William's son, Harley Procter, a scientist by training, and James Gamble were known to run experiments in the plant to improve product performance. Ivory just may not have been an accident.

Hans Rosling was a very famous Swedish physician and statistician, who illuminated facts revealed by deep studies of historical data, particularly on health. Quite surprisingly, just some ten years ago following a major reflective study, he concluded that **IVORY soap was clearly a major "technology" that changed the world, because it had such an impact on public health at the time.**

Over the years, the Company developed products using its technical strengths and outstanding people to create many innovative new brands, and with effective acquisitions it continued to grow. Today, it operates in 10 different product segments with 65 brands.

| CATEGORY | MAJOR BRANDS |
|---|---|
| BABY CARE | BABY CARE |
| FABRIC CARE | TIDE & ARIEL DETERGENTS & DOWNY/ LENOR |
| FAMILY CARE | BOUNTY TOWELS |
| FEMININE CARE | ALWAYS |
| GROOMING | GILLETTE & VENUS |
| HAIR CARE | PANTENE, HEAD & SHOULDERS |
| HOME CARE | FEBREZE…SWIFFER … DAWN…CASCADE |
| ORAL CARE | CREST & ORAL B DENTAL PRODUCTS |
| PERSONAL HEALTH CARE | VICKS COLD PRODUCTS & ZzzQUIL |
| SKIN & PERSONAL CARE | OLAY SKIN PRODUCTS |

P&G evolved from an American to a global Company beginning in 1930 with the acquisition of the Thomas Hedley Company in Newcastle, England. Expansion occurred over the years, mostly as an exporter, but geographical expansion became a highly focused growth strategy during the 1970-2000 period. Today, P&G is a well recognized global Company with operations in more than 70 countries.

The original focus on superior performing products with outstanding consumer recognition and acceptance remains the cornerstone of its strategy. The Company's focus is having the #1 brand in every category in which it competes. A superior performing product coupled with outstanding advertising and marketing of the product's strengths drove the Company's success.

To achieve the desired product superiority, the Company continually invested in internal Research & Development. This always included a Corporate R&D investment that allowed the exploration of technologies and stretching, longer-term product goals that formed the basis for almost every major new brand.

I studied biochemical engineering at the University of Wisconsin. Biochemical engineering was a new curriculum in 1965 when I started. Today it is the prominent focus in chemical engineering programs across the world. Graduating chemical engineers at that time were primarily expected to go into chemical, energy or environmental industries. However, to the disappointment of my professors, I happened to be interested in retail or commercial products. I always was a "tinkerer," taking things apart, building forts, bridges over streams, fixing household physical problems, and enamored by automotive concept cars whose development was well off into the future. The idea of working on the development of new and better consumer products was of great interest.

I interviewed many different Companies as I sought to line up employment after graduation - food, beverage, beer, alcohol products, fragrance, paper, cosmetic, personal care. Jobs were plentiful at the time with the Vietnam War and the Space Program capturing a large percentage of the young engineering workforce.

I had an interview trip to P&G. My expectations were low. But during the visit, the exposure to the broad range of innovative products, the desire to be the best, working with high-achieving people, and the exciting stretching work going on captured my interest. I was sold. I often referred to the feeling I had then as loving:

## "the smell of the place"

I started in early 1961, and began as a process development engineer in the P&G Food Division. It was the start of my 40 year career, which formed most of the learnings and experiences I will share in this book.

# SECTION ONE

## "THE ENVIRONMENT FOR PRODUCT INNOVATION"

# 3.

# INNOVATION HAS TO BE LED AT TOP OF COMPANY

To achieve a truly innovative company the thrust behind product innovation has to be led from the top - the CEO. Innovation is very difficult and organizations don't naturally want to take on the risks associated with stretching programs. Most individuals don't really want to take on highly challenging objectives that might delay achieving career advancement. **The CEO, therefore, has to demand innovation from the organization**, must be personally involved, and - of course - support it financially. The organization has to be convinced that developing major, new and disruptive innovative products is EXTREMELY important to the Company's leadership, and to the Company's success. Every Company wants innovation, but most give

## ""THE CEO HAS TO DEMAND INNOVATION FROM THE ORGANIZATION"

it "lip service" and leadership for innovation gets delegated and diluted in the organization.

Procter & Gamble's Mission Statement marked the criticality of superior products:

> *"We will provide branded products and services of SUPERIOR quality and value that improve the Lives of the World's Consumers, now and for generations to come"*

The CEO's job on delivering those SUPERIOR products and the need for a superior R&D organization to deliver the innovation is strongly implied, and the key premise behind this book.

Working for the four different CEO'S as I did provided a spectrum of different experiences. They were all effective, but with significantly different approaches.

All the CEO's I had the privilege of reporting to were outstanding business leaders. Their accomplishments are well documented in financial reports, many other books and publications. I won't try to capture those here. Instead, I will outline the key management approaches each brought to the product-innovation and management process during the years I reported to them.

 JOHN SMALE - (1981 - 1990)....I interfaced with John continually over my career, and reported directly to him in 1987 when I became CTO. John was probably the greatest advocate of product innovation since the Company's founders. Shortly after becoming CEO, the editor of "Moonbeams", the Company internal magazine,

interviewed John. He talked extensively about the importance of P&G"s most innovative brands, and surprisingly commented:

**"We may be known as a marketing Company, but we are fundamentally a Research & Development Company".**

He supported this statement when he promoted 6 R&D leaders to Vice President and members of the Executive Committee. A few years later, and for the first time, he promoted an internal R&D manager to the Board of Directors. His goal was clear - give R&D a full seat at the table.

In his last talk to an audience of the senior P&G managers in 1998 he outlined his "Principles for Success". Number one was INNOVATION. …

**"Nothing else we do, Nothing…..good Advertising, lower costs, better distribution, financial planning can substitute for product innovation.**

Subsequent chapters will cover my interactions and key learnings working with John Smale. But there is no doubt that he was a standard bearer for leading product innovation as a CEO.

 ED ARTZT - (1991-1994)…..Ed was a very tough, no-nonsense manager, who was always very supportive of the R&D effort. I had worked under him as a young engineer in Foods, when he was General Manager, and as a Category R&D Director in Europe when he was in charge.

Different than John Smale, Ed chose not to be highly involved in the product evolution process…He had great trust in the R&D work…

However, he was actively involved in new product rollouts, keenly focused on effective execution when the time arose. A key example was the global launch of Pantene hair products (Chapter 26). The shampoo featured the breakthrough 2-in-1 shampoo with conditioner. Ed drove the advertising and package design elements from different P&G country organizations outside the US, and then proclaimed, "We are going to get this fully global within a year - nobody is going to take this away from us"....This was more than 100% faster than we had ever expanded a brand globally. There were huge issues in manufacturing, ingredient supply, regulatory clearance and, advertising copy in multiple languages.....But the blitz approach was successful. PANTENE became the #1 global hair care brand.

His trust in R&D was really put to the test when he boldly challenged Unilever publicly in 1994 (Chapter 22). Unilever's Persil Power detergent contained a whitening catalyst that deteriorated colored fabrics It was a terrible safety error that they had made in a laundry product formulation. There was a carryover effect which also had an impact on our product's performance. It was a one-of-a-kind encounter that only a super-strong personality like Ed could have taken on.

 JOHN PEPPER - (1995-1998)......John stands out as the "soul" of Procter & Gamble. He always was a super-principled, people-focused, do "the right thing" leader. I worked under him during the launch of Bounce; in the Euro-product effort in Brussels; and in the launch of Liquid Tide & Tide with Bleach. He was always a great supporter of R&D, and when he addressed researchers, they went away inspired and energized by his avuncular demeanor showing extremely deep and sincere interest in their work.

He was CEO when we considered acquiring TAMPAX to add to Always, our very successful feminine protection brand. This was a huge challenge for P&G, having removed our RELY Tampons product from the market a decade earlier amidst concern and uncertainty about the tampon's effect - or any tampon's effect - on Toxic Shock Syndrome disease. He challenged me to develop a position on the subject. Much had been learned since the Rely experience, but John's highest-level concern was for the Company's reputation which overshadowed everything else. We ultimately concluded it was safe to acquire TAMPAX. We satisfied John and the Board's concern, acquired the brand, and have since grown and globalized TAMPAX

John was locked into the importance of superior products. In an in-depth interview in recent years, following his point about products in the Company's "MISSION STATEMENT", he was asked, "What would happen if creating superior products was no longer a key priority?"

> **..."Whoa!...I'd say we would be DEAD in the water. A lot of people would end up leaving the Company, because it no longer would be the company they know.....I'd sell my stock.....I'd get out quick."**

DURK JAGER - (1999-2000)....Durk was clearly the most "hands-on....do everything possible"....and "do it fast" leader of product innovation. I had worked with him when he was a country general manager in Europe;,when he led and turned-around our Japan business, and when he was in charge of our US business as COO and President in 1995. From the start of our relationship he wanted to eliminate any barriers and bureaucracy that impeded developing big, new products. He brought his own new ideas, and supported every disruptive product idea we had conceptualized.

We started a very unique internal New Venture program (termed ILT), in which he and I shared leadership. The results were a string of successful new brands (Chapter 27), including the iconic SWIFFER and FEBREZE brands. We evolved to an"Open Innovation" strategy which involved opening our patent portfolio for sale or license, and launching new partnerships with outside suppliers and manufacturers known for their strong but differentiated technical strengths. Further, we challenged the organization to increase connections outside the Company to access desirable technology.

Our active relationship on innovation was very close, and I valued his full support and personal involvement.. The Company's innovative spirit was at an all-time high, and great business results were to follow in subsequent years behind the venture product launches.

# 4.

# CTO NEEDS TO REPORT TO THE CEO

The relationship between the CEO, and the Chief Technology Officer (CTO) or Chief Research & Development Officer is absolutely CRITICAL. I use the title CTO in the book, because that's what my title was as head of Research & Development. I realize today that the title CTO may most often be used for Chief Information Technology Officer. The CTO or Chief Research & Development Officer should report directly to the CEO. The CTO is essential to the success of the upstream innovation program. He/she must be empowered - must have the full support and attention of the CEO. This strong and direct relationship between the CEO and CTO was a part of P&G's structure for most of its entire history. Even before there was a formal R&D organization and a formal head of the organization, the CEO would reach out to key upstream R&D leaders to gain awareness of big ideas that were being championed within that group.

I had the chance to talk with Vic Mills, probably the Company's most prolific inventor and innovator, whose career spanned 35 years starting in 1926. He and his team were behind inventions on soap and oil processing, CRISCO shortening, TIDE detergent, the original design of PAMPERS, the patent behind roller milling cake mixes which became the engine behind DUNCAN HINES baking mixes, PRINGLES, among others. I asked him how he interacted with Company management?. He said, …"We had complete freedom….. kinda did our own things… we liked to think big….time would go along, and I would get a call to come downtown and talk. I would fill them in on what was going on…..there would be a lot of interest, and pretty soon, new people came on the scene, and the ideas would be moving forward."

That seemed ideal, but normally every major, big, new idea at the point of conception has only small support, with the majority believing that the idea will not work for a variety of different reasons. This was true throughout P&G's history, throughout my career, and is the case in all corporate situations.

Machiavelli, the Italian philosopher said,

# There is nothing more difficult to plan, ....more doubtful of success.....nor more dangerous to manage.... than the creation of something entirely new...

**..The innovator has enemies of all who profit by the old way or only see the risks….and merely lukewarm defenders of those who would gain by the new or who primarily fear any accountability.**

Most business managers see their job as one of judging and challenging - rather than encouraging and supporting. So, without fail, the supporters of a big idea at the early stage will be a small minority. This has been the case of **EVERY** major success P&G has had, or major idea I have studied in other Company situations.

The Austrian Economist Joseph Schumpeter described innovation and entrepreneurship as "creative destruction". Successful companies have learned how to operate as an ambidextrous organization. That is, having the ability to support projects designed to incrementally improve the performance and profitability of an existing business, while simultaneously supporting projects to obsolete it.

Once a disruptive new product idea is conceptualized it becomes the target for the establishment to attack. Finance challenges the sales projections and the discounted cash flow as unrealistic. Engineering claims they will have to invent process equipment to make the product at an acceptable speed. Manufacturing announces the staffing level will be 50% higher than proposed. And Regulatory has grave concerns about the safety of one of the key ingredients.

This part of the new product development cycle is affectionately known as the

**"Valley of Death",**

the phase in a new product's development between the euphoria of conceptualization and ultimate commercialization. It was the subject of an entire book. "Traversing the Valley of Death" *(Paul Mugge and Stephen Markham Center for Innovation Management)* It's the place where resources are inadequate, funding is drying up and product champions become disillusioned and broadly frustrated. It's where many good product ideas go to die.

Therefore, having a strong CTO who can identify and shepherd important ideas through early vulnerable stages is essential. Further, this individual has to drive home the standards and Corporate goals to the entire R&D organization, and hold them accountable. In turn, the CTO has to be held accountable for delivering the necessary technical innovations to continue to advance and grow the Company. Needless to say, having the right leader as the CTO, just as with the CEO, is very important. But to have any chance for the product innovation stretching goals to be accomplished, the CTO has to have a "seat at the table". Among many other reasons, he/she is very often trying to champion business decisions which are at odds with the majority point of view.

In my first meeting with John Smale, after I was appointed CTO, John said, "**Outside of geographical expansion, the Company has major growth only through major new brands, or through acquisition. Clearly growing major new brands internally through the superior product route is the most financially favorable. I'M COUNTING ON YOU FOR THIS.** There was little doubt what his expectations of me and the R&D organization were, and it remained top of mind as I went forward with my work.

# 5.

# "THE NECESSITY OF BEING OPEN TO INNOVATION, CHANGE, AND NEW IDEAS"

We all realize that not all companies are interested in change or continual improvement. They cling to what made them successful until their business is disrupted by technological, marketing or distribution changes. Then they are in survival mode.

 We all know the "buggy whip" story. A Buggy-Whip company driving to make better buggy whips when the need for their product was disappearing. In 1890, there were some 13,000 companies involved in the wagon/carriage business. With the advent of automobiles, "change mode" was a must. Companies making buggy-whips and carriages, both of whom had unique and deep capabilities, did not change and went out of business. Companies making roller bearings (Timken), axles, and carriage lamps pivoted their focus, grew, and thrived. Being open to and grasping change was essential.

I had a real-life experience during my time at the University of Wisconsin that really drove home this point.

In our studies in biochemical engineering, we did a great deal of experimentation on continuous fermentation, as well as the processing effects on yeast, molds and bacteria. At the time, it was very "leading edge" work in academic biochemical research.

Being in the heart of the "beer world", some fifty miles from Milwaukee, I was excited about the possibilities of using these studies to make big inroads into the beer industry.

At the time, the Schlitz brewing company was neck and neck with Anheuser Busch for the leading brand position. Pabst Brewing, which had just acquired Blatz Brewing , another Milwaukee competitor, rounded out the top three. Miller Brewing was also a major player as #4.

These companies had all been founded around **1850** and located in Milwaukee, which was very close to outstanding grain producing areas. All these companies were very innovative in their beginning with major advances in quality control that resulted in a consistent product. The quality of their brewing formulas created great consumer appeal, reminiscent of the great European brews from Germany, Austria , the UK and the Czech Republic.

Subsequent innovations included major advances in brown bottles, aluminum cans, and continually orchestrating processing changes to provide very efficient production. Schlitz had been the key factor in establishing Milwaukee as the brewing capital with its advertising, "The Beer that Made Milwaukee Famous."

I eagerly went through campus interviews and had on-site visits in Milwaukee with Schlitz, Pabst and Miller. I had great anticipation in describing our landmark work in continuous fermentation. Although the interview visits were of high interest, I was extremely disappointed with the results. Although I was quite eager to talk about the different avenues of study that I had been involved in and how it might apply to their business, both the technical as well as the business people with whom I talked seemed completely uninterested. When I talked about their needs for engineers and particularly biochemical engineers, there was no excitement or interest in what seemed to be a significant improvement to their brewing process. I was stunned.

When I compared these visits to later ones with other chemical-based companies, like Procter & Gamble and Dow Chemical, the difference was like "chalk and cheese."

In the following years, Schlitz ran into MAJOR quality problems. They had adopted a strategy against Anheuser Busch of "If we can't out-produce 'em, we'll beat 'em with greater profits." They went through a period of changes in their processing and utilizing less expensive ingredients, losing sight of the taste character critical to Schlitz users. Beleaguered with customer complaints and bad press, their business spiraled downward. In the late 60s they were faced with having to recall ten million bottles of beer from the trade. Customers revolted and shifted their beer purchases to Anheuser's Budweiser. Shortly thereafter, Schlitz essentially went out of business, going from number one to an insignificant player in the beer business, all in in less than fifteen years. The brand was purchased by Stroh's and then became a minor brand at Pabst.

Encouragingly some 15 years later, Miller led the introduction of Light beers, a major discontinuity in the beer industry. This was a major strengthening factor for Miller, and again demonstrates the criticality of product innovation.

 Continuous beer brewing, which I had spent time on in the university, was adopted first in 1958 by the New Zealand Brewing, Ltd. company, who built their structure from the start with this latest fermentation technology. It proved to be very successful in establishing the number one beer in New Zealand - TUI.

Whether any ideas that I had in mind would have been helpful to these Milwaukee beer companies, of course, is completely unknown. However, the fact that these companies showed almost no interest in considering change was certainly not inspiring to a prospective employee, nor provided any route that might have benefited maintaining their leading position.

Being resistant to change for any product-producing company is deadly. Deadly for the business and for acquiring and motivating top employees.

# 6.

# "KEEP INNOVATING OR FADE AWAY"

One doesn't have to look into history very deeply or very broadly to quickly understand that once an innovative product is launched, it's just a start on the innovation needs for that product.

History has a myriad of once-unique, creative products that pass from the scene as competitors overtake them with better ideas or performance or as consumer habits and interests change. We have an example within P&G that vividly illustrates the importance of continued innovation.

The story involves two iconic consumer brands - IVORY bar soap and TIDE laundry detergent.

IVORY was P&G's first major brand. It was introduced in 1879. It was a superior mildness bar soap, absent of dyes and heavy perfumes, and the only one that floated. It was positioned as 99 44/100% pure, and sold at a premium price. It became the leading bar soap.

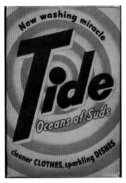

TIDE was the first highly innovative laundry detergent, and was introduced into the US in 1946. It contained phosphate to counter water hardness, new oil-based surfactant for soil removal and suspension, and spray-drying for easy dissolving. It became an immediate success.

The share progress of these two brands over the next 75 years is shown in the EXHIBIT.

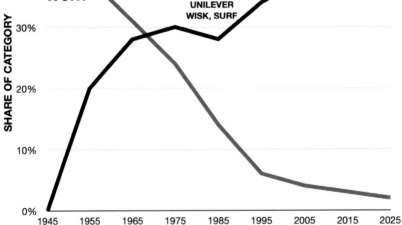

## Major Innovative Product History

TIDE never stopped being the best performing and most preferred product in the market with some 100 significant changes and innovations over the years. The addition of better surfactants, the replacement of phosphate for improved environmental effect, highly effective enzymes, soil suspending agents, compaction for better convenience, more convenient packaging and sizing, a breakthrough liquid form, a unit-dose pod, and countless more improvements and changes. It always was able to command premium pricing in the category.

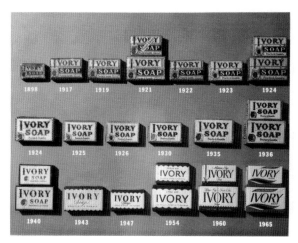

By comparison, IVORY bar soap did not maintain its competitive edge over its 140+ year history. Competitive products were introduced with added moisturizers and oil-based surfactants, like Unilever's DOVE. These products were milder, with more, preferred lather, and avoided forming "curd-like" residue in hard water areas of the country.

IVORY was viewed as a classic, iconic brand. There was great company leadership resistance to make any change, particularly after NEW COCA-COLA, also an iconic brand, stumbled trying to make a basic flavor change in 1985. Fortunately for

them, they were able to restore Original Coke to the market without a significant loss of market share.

A modified IVORY bar soap product with superior acceptance to original IVORY was test marketed in the 1970's at a premium price. Of course, it was no longer 99 44/100 % pure soap. It successfully built IVORY share, but was not expanded, or introduced as a flanker alternative. Despite further reducing the cost of IVORY to consumers, IVORY share continued down.

IVORY was still viewed by consumers as a wonderful brand, mostly from their childhood experience, and I remember in one study, the predominant attitude toward IVORY was "A great brand, I think my neighbor would really like it"

So, over the years, IVORY went from a premium-priced, superior mildness, unique product to a lower-priced product than key competition, and lost consumer interest. It's hard for me to imagine why there wasn't some satisfactory product answer to leverage IVORY'S iconic image to maintain the brand - but no one came up with it.

Whether IVORY should have been upgraded is very difficult to judge, as the company's reputation was closely intertwined......But one thing is clear, without change, the brand was not able to keep pace with the market.

# 7.

# MUST BE CLOSE LINKAGE BETWEEN R&D AND DECISION MAKERS

I had started at Procter and Gamble in the Food Division R & D organization as a process engineer. I was about two years into my launch at P&G when I was assigned to the F49 project. F49 was to develop a breakthrough margarine product. One with consumer superiority to the global market leader, the Unilever brand, Imperial.

Procter and Gamble, of course, had been very basic in vegetable oil processing with leading brands Crisco and Crisco Oil on the market and significant oil processing capability in several plants. The thought was we could build off of that base and develop a product in the margarine area. Margarine  was a very fast-growing category. Butter was under attack for heart health issues as well as, of course, much higher cost. Margarine had good spreading capability and sold at a reasonable cost, but existing products lacked the kind of butter-eating quality and butter flavor that consumers desired.

Margarine in the US was made under a FDA standard of identity. This meant that there were tight controls on all the ingredients that could be used in a margarine product.

Procter and Gamble had a patented idea to build on. A process for delivering the desirable mouth-melt characteristics of butter had been invented by a food division employee, Dr. Robert Dobson. It was called a double emulsion process. That is ,there was an oil and water emulsion made which was then incorpo-  rated into a water and oil emulsion. This configuration gave a very positive butter-like eating quality. A product development team was put together which had three main engineers who were working on the project. Cliff Larsen was in charge of the process for making the product, a continuous process which incorporated the double emulsion technology. Bill James was the engineer in charge of the oil selection. Needed here was the leading-edge composition supported by health authorities for most desirable heart health, as well as one providing the necessary in-use stability and spreading characteristics. I was given the job of developing the butter flavor. The standard of

 identity made it very difficult to create a butter flavor. There were only two ingredients that were allowed to be added artificially: diacetyl and delta lactones. Both are natural flavor compound in butter. These compounds were included at the start of the US standard of identity formation. However, these ingredients only allow the achievement of a very marginal quality butter flavor in a margarine product.

Gas liquid chromatography was at an early stage of development in the early 1960s, but we used this technique to go through a process

of trying to identify additional flavor compounds from butter that would be most useful in our product.

Through a combination of identifying the compounds and then trial in products I was able to zero in on several compounds which I felt would give us a terrific improvement in butter flavor if we could develop them naturally. It was a combination of short chain fatty acids, a selection of methyl ketones, and dimethyl sulfide. I felt there was a chance by using cheese making techniques along with enzyme treatment of butter oil that we could generate enough of the desirable compounds. I zeroed in on a bacterial fermentation of milk as well as a fungal fermentation, and also lipase enzyme treatment of butter oil to create free fatty acids. Through experimenting with the fermentations and enzyme treatment, analyzing the results and then blending the products we found we could produce a very desirable butter flavor. These processes and their combination actually ended up in a patent application and approval.

At that time, paired comparison testing, that is, the testing of your product against the competitor in a blind labeled product situation, was the preferred way to prove the quality of your product. We prepared for the preparation of our product for the test, which was a challenging scale-up for us. We finally got everything to go smoothly, the product was prepared and we placed it  into a national paired comparison blind test. The goal was a significant advantage, which would be at least a 56-44 result amongst consumers' choices on the large consumer base that we placed.

The results finally came in. It was a spectacular result - 70-30 preference by consumers. A HUGE blind test win by Procter and Gamble standards. We were very excited. We now were in a situation where we could move into a test market. The launch proposal was prepared, the plant for the production for the test market was identified, and we were ready to go. The plan went forward for approval in the Company, and we stood ready to get started. About a week later we got the news and I'll never forget how devastating that it was. We were told that the company decided NOT to proceed. The company's top management had decided that they did not want to get into the refrigerated food business.

How could this happen? Why wouldn't you give this kind of product a chance? Why did we find out now about the company's view of refrigerated foods?

All the knowledge we had built. The patents we had established. The product result we had achieved, and of course, the tremendous amount of work and effort that we and the people working with us had put into this project. **ALL DOWN THE DRAIN**!.

Bill James and I were reassigned. Cliff Larson actually left the Company. The great, one-of-a-kind pilot plant that had been created was dismantled. The patents were put on a shelf. Nothing was gained from a tremendous effort that had met its goal.

It changed how I viewed product development in a major way. I said to myself, this can never happen again. I looked for a career path internally at Procter and Gamble that would always give me a chance to be at the interface with the business community, where I could be sure on the projects I was involved that this would never happen again.

One would normally have said that this could NEVER happen at Procter and Gamble. Well, it shouldn't have, but it did. And because I had many interactions with other company R&D heads over the years, I knew it was not an entirely unique situation. We learned the hard way how important it is when developing a major innovation to have full agreement on the project from the decision makers in the company. Otherwise, there is high risk for an absolute waste of time, money, effort, proprietary assets, and people's energy.

# 8.

# "WHAT'S THE BEST STRUCTURE FOR MAJOR NEW PRODUCT DEVELOPMENT"

Developing major new brands, which are so critical to the growth of a company, was a major challenge at P&G and in all company structures.

As I reflected on the situation at P&G, I could see where the company was most successful over the entire history of the Company in developing major new products. These products became new brands, or major additional products as part of an existing brand that greatly expanded the category.

From the start of the company in 1837 all the way through the late 1950s, (over 120 years), the R&D work was done corporately, and basically reported to the top of the company. The line of communication was from the idea generation and development group right to the decision makers. Decisions could promptly be made to move ideas forward and the process worked extremely well. We took on acquisitions, and used those new technologies along with our existing technologies to move into new products like vegetable oils, detergents, cake mixes, paper products, Pampers diapers, coffee & nut roasting, Pringles potato chips, etc.

The next major advance was in the mid-1960's, when CEO Howard Morgens, and his eventual successor Ed Harness, laser-focused on the "Blow-through Drying" invention for Paper-making. It put P&G into a viable paper business, with BOUNTY, CHARMIN, and PUFFS.

The next segment of major advances happened in the **1980**s. Key here was CEO John Smale's great focus on big, new products, and to his approach for regular high-level product meetings with each of the R&D directors of our various businesses. This put the CEO in direct contact to the big ideas, and led to rapid advancement of major new products and brands in the Company. The list is extensive…ALWAYS DRY-WEAVE feminine hygiene products, PAMPERS ULTRA-THIN diapers, CREST TARTAR CONTROL toothpaste, TIDE LIQUID, TIDE with BLEACH, PANTENE shampoo, the MICRO-TEXTURED paper technology which propelled BOUNTY TOWELS & CHARMIN BATH TISSUE, and the high-risk decision for starting an 18 year major investment in a new drug for osteoporosis (ACTONEL).

The third major segment of major new brand development happened in the late **1990s** with the advent of the new Internal Venture program. Here the CEO , Durk Jager and I, as CTO, launched and were in direct contact with major new ideas. Decisions on staffing, investment and moving ahead could all be made very quickly and avoided any significant delays. Coming from that were SWIFFER, FEBREZE, CREST WHITE-STRIPS, GLAD PRESS & SEAL wrap and GLAD ODOR CONTROL AND FORCEFLEX BAGS, THERMACARE DISPOSABLE PAIN RELIEF PADS, ALIGN probiotic and several others.

When you look outside the company, you see a common occurrence of major new products and brands being developed with the business leader intimately tied with the development. Situations like Henry Ford on Ford Motors, Steve Jobs on Apple, Elon Musk on Tesla, Ray

Croc with McDonald's products and service, John Chambers on Cisco Routers and Systems, Larry Page & Sergey Brin on Google, Howard Schultz on Starbucks, the Harley & Davidson brothers on their motorcycle, and on and on.

The conclusion I draw from all this is that;

## "if one wants to develop major NEW products and brands, the best structure is to have the development work that's underway tied directly to the company's decision makers

- at least until the product is on its way. That format seems to have worked throughout the one hundred and eighty years of Procter and Gamble's history, and certainly fits my career experience entirely. It may seem radical to most company structures, but in fact can evolve very straightforwardly. It is sure to be efficient, although it takes tremendous commitment on the part of the CEO, and a very engaged Chief R&D Officer.

Interestingly, as I was drawing to a close in writing this book, a Wall Street Journal article appeared (7/26/2023). It's a story of one of P&G's oldest and strongest competitors - Unilever. They had just announced their quarterly results, which were positive on profit and sales achievement. However, newly appointed CEO Hein Schumacher voiced concern over falling share positions. He indicated  his immediate personal focus was to directly interface with the R&D organizations to achieve a "high quality, focused and science-backed innovation pipeline"…."Our brands should be winning superiority tests week in and week out. When they aren't, we should be taking decisive action".

It seems like Mr. Schumacher also understands the critical value of the close CEO interface with the R&D organization to foster product innovation.

Perhaps one of our book drafts was leaked? 😀😎

# 9.

# "HOW MUCH SHOULD YOU INVEST IN R&D?"

The issue of the investment in R&D is a constant question in most companies. One thing can be sure…..No R&D manager would ever feel that they have sufficient funds for all he or she wants to attack. No R&D budget is big enough. But, in the end, a decision gets made, and the R&D program is adjusted to be handled within the budget constraints.

**No R&D budget is big enough.**

History, of course, is a great teacher. One can learn a lot by the amount of investment that had been required over different points of time in a specific category. How it needed to grow with added brands, added geographies, new relevant competition, specific crisis situations involving environmental, safety, or legal challenges, or competitive shortcomings in performance. Or alternatively, situations where investment would want to be reduced based on strategic importance of the category, or the sale, maturity, or financial importance of brands.

As one would expect, the robustness of the innovative plan for a given category was critical to determine an investment decision. For a particular highly demanding innovation project, it would be necessary to make an unusual new investment. Likewise, a weak innovation plan would obviously urge challenging the plan, and/or forcing an investment reduction.

Some categories, of course, would have ample financial strength to do whatever they wanted. Others did not have the financial ability to take on a project critical to "changing the game" for their business. As CTO, I could intervene and help make sure that our investment was put against our best opportunities, as opposed to who could afford them. This often led to heated discussion, and a level of unhappiness with decisions. It was always easier for a general manager to favor supporting existing brands versus investing in work to obsolete them. However, I can remember many situations where such forcing led to major advancements. An example was the shutting down of work on existing household cleaning products to work exclusively on developing SWIFFER, which changed the game for the category.

Many times the amount of investment can go up and down with the economy. My approach was to try to keep investment as constant as possible. I took the position with the various CEOs that I worked with that we should invest as a percent of sales with our best competitor, or at least at the same total level of total investment, and be judged on our ability to compete against them.

This happened to be Kimberly-Clark in paper and diapers, Unilever , Henkel and Kao in laundry and cleaning, Unilever and L'Oreal in beauty care, and Colgate in Oral Care. Of course, the percent of sales in R&D investment will vary substantially from business to business. For example, food was always a lower level of investment, primarily because the infrastructure to compete in this business is so much less than it is in some of the more capital intensive businesses we were in.

There are some businesses where you just cannot afford to be behind in product acceptance.

The best example I know is diapers. In the diaper area, your customer has only a two year span of interest, and so consumers are constantly turning over in the market and always looking for the best product. If you fall behind in diaper design your business will suffer, and the financial penalty of that far outweighs whatever level of investment is absolutely needed to keep product superiority.

Of course, there are situations where, as a company, you're making a major investment to NEWLY enter a particular category. That was the case in pharmaceuticals where we had a relatively small business that we acquired, but we put major effort into the development of a new drug for osteoporosis. At the peak of investment, we were investing something on the order of 40% of sales. At the peak point, we had $600 million dollars into just the final clinical for gaining FDA approval. These were very challenging times, and periodic update reports to the board always turned out to be very stressful.

# SECTION TWO

## "INNOVATORS MUST BE CHAMPIONED AND REWARDED"

II

# 10.

# CHARACTERISTICS OF AN INNOVATOR

Over the many years I've been able to watch great innovators, and see great innovations evolve both inside and outside the company, I will try to frame the common characteristics I have observed.

Great innovators seem to have these common characteristics:

## "MASTERY, PASSION, CONNECTEDNESS, FREEDOM, PERSISTENCE & STRETCH

…..and they seek a supportive system in pursuing their careers.

**MASTERY** is a self-evident characteristic. . There is a laser focus on a given area of science or technology. But the innovator also has the rare ability to understand and connect a broad range of scientific areas. The great innovators never stop learning, but thrive on expanding and sharpening their already keen capabilities. They are keenly aware and acknowledge the limits of their own knowledge. They try…they fail….they **learn**…and go on. Elon Musk declared: "If things are not failing, you are not innovating enough"

**PASSION** relates to strong internal motivation. It's that dedication to first understand and define the problem…then solve the challenge.

The reward is to be able to select and solve the next challenge. History shows that the innovator is a serial problem solver, and they commonly have more than one big hit.

There is no such thing as a commodity or the perfect product. There are only situations demanding a major change. Many times in the consumer product world the goal can seem mundane. How can one get passionate about solving underarm odor, or effectively removing soil from a floor, or keeping babies dry? That's where it requires passion. No problem in front of an important goal is too insignificant.

**CONNECTEDNESS** is the ability to use other people's experiences to solve problems. The innovator is stimulated by new learnings. Consistently curious….How happen?…What if?…. Some act as "Gate Keepers", bringing in and distributing tacit knowledge to colleagues. They actively reach out to gain perspective, thereby increasing their knowledge base. They are connected inside and outside the Company. They are active members of several Communities of Practice.

**FREEDOM** frees up the mind to focus on what's important. It often requires the strength and confidence to do the "unexpected" thing. Freedom is highly valued by the best innovators, often…the most important factor in their work environment. We once did a study among our Research Fellows regarding improving R&D productivity.

# "GIVE US LARGE BLOCKS OF UNINTERRUPTED TIME TO WORK AT THE BENCH"

was the dominant feed-back. (WMJ article IRI June, 2003). There's little doubt that the transaction costs of global, multi-discipline teams, and frequent management reviews take too much time from experimenting and inventing - and most of it is unproductive for the key innovators.

Toan Trinh, the inventor behind FEBREZE, is the holder of 220 patents, the highest ever within P&G, and the first P&G researcher named as a "Distinguished Corporate Inventor" by the National Invention Hall of Fame. He provided his view on the biggest barriers to innovation. "**People don't have time to think. For a long time, my pattern was to spend 30% of my time just thinking.** In recent years, first with voice messages, then email and finally with texting, I found myself feeling obligated to spend most of my time answering long lists of questions. Good for managers, not for people doing the work ".

This is the area where research managers can make a big difference. They need to create an open, entrepreneurial environment that allows people to pursue their big ideas without undue bureaucracy. In this way actual work time is maximized. Ideas are challenged, but respected. The opportunity exists to try big, and fail in a dignified way When failure happens - and it will - there is always new learning that makes the next attempt easier. It's a bit like mountain climbing . Seth Godin, a very accomplished author, once said: "Big ideas are little ideas that no-one killed too soon."

**PERSISTENCE** is, of course, essential for major innovation. The technical challenges are very difficult. They don't lead to great comfort nor an easy path forward. They never generate full alignment with colleagues at the early stage. That's been true of every major innovation in history.

Leading innovators understand that they will be in a minority at the early stages. It's lonely but that's okay with them. They will shoulder the uncertainty and criticism of not being "team players". They remain humble and open to critique from supporters and skeptics. They adapt positively to diversity. It's their persistence that keeps the project alive. They know that.

I mentioned earlier about "THE VALLEY OF DEATH" - where EVERY major project faces a very strong challenge to stop - for countless possible reasons. If stopping is not the right course, people have to "step up". Persistence gets the supreme test. You will see that almost every great innovation covered in this book came to this "VALLEY OF DEATH"- a critical juncture! It was the PERSISTENCE of the project champion that fortunately won the day. It could have been the CTO, a R&D manager, but also many times it required the persistent researcher.

**STRETCH** is, of course essential for major innovations. It's not the kind of thing that generates much comfort. In fact, it does the opposite in creating a great deal of personal stress. Whereas most individuals seek the minimum goal for being rewarded, great innovators reach for "game-changers". It never generates full alignment, particularly at the early stage. That's just a part of every major innovation in history.

I have been involved in many major brand innovations that I worked on personally. Without fail, the majority of people around me believed it would not work. But I persisted promoting the product idea that I had. Eventually the strength of my conviction convinced the management to give it a shot That's all one can really expect.

If those are the characteristics of an innovator....is it possible to identify them at an earlier stage of their careers? We once tried to take a look at the backgrounds of our 10 highest achieving, proven innovators. There was no way to soundly identify characteristics…but it was amazing to us how many similarities existed.

As youngsters, they tended to "work with their hands"…farmers, construction, mechanics, manufacturing.-both male and female. They were "do-it-yourselfers" throughout their lives. Generally found the answer themselves to everyday problems - household repairs, tractor and automobile repairs, 'fix up the old, rather than buy the new". Experimenting in the kitchen and redesigning older clothing items. We concluded they had a long history of tackling the challenge in front of them….and it had continued through their academic studies on into their work careers.

I often would ask recruits and new hires questions that would try to elicit the traits of budding innovators. Scientifically unjustified, but I would always spark to an individual with one of these self-directed backgrounds based on these observations.

A question may be what about the best innovators being risk takers? Well, none of the best inventors, and most successful discontinuous thinkers I know would ever think of themselves as taking high risk. They are very conscious about risk and, in fact, they are very concerned about keeping risk "low". They realize that stretching aggressively at the earliest stages of investigation is normally a very reasonable cost proposition. There usually isn't a lot at stake financially, and the potential learning from the stretching is very high.

Most inventors are not entrepreneurs, and vice-versa. It's kind of like a fisherman and a chef. The fisherman is trying to catch the elusive fish, and the chef is looking to apply the Epicurean expression and bring it to the customer for a profitable return.

As I think about the internal venture program at P&G (Chapter 27), it was the ultimate solution of bringing inventors and entrepreneurs together. Ideas with big potential were brought to the Company decision makers to, hopefully, get their approval to move ahead. Seems like a pretty good situation, doesn't it?

# 11.

# "INNOVATOR RECOGNITION SOCIETY & RESEARCHER CAREER PATH"

Everyone would agree that the most critical element of an effective innovative system is the quality of the people. At P&G we were committed to hiring the most accomplished and finest engineers and scientists. During most of my career we had no specific R&D reward system. R&D people's accomplishments were evaluated as they were throughout the company, and people were paid and promoted based on their performance. The intent was to operate as a meritocracy. However, the only way to get visible recognition internally in the Company was to become a manager and advance through various levels with promotions. So for managers, status was easy to discern - it came with a bigger office, title, and internal announcement of your promotion.

Outstanding technologists were the critical capability we needed. But if a technologist wanted to remain "at the bench" focused on doing scientific or engineering work, you might be well paid, but you would have to be satisfied with the challenge of doing exciting and challenging work. There was no visible recognition of what you accomplished. **However,**

# "you couldn't wear your paycheck on your forehead"

The system actually drove technologists away from their technical development work, which was the prime work needed in the organization. Often good technologists became poor managers. A real Catch 22!

"People are motivated by more than just take-home pay: a passion for a higher purpose: the desire to be recognized and appreciated: the ambition to grow one's skills".. Bahcall, LOONSHOTS (2019)

I knew what we had was not satisfactory, but the track record of many other company's dual ladders was not very inspiring. The technologist ladder, instead of being inspiring for a technologist career, often became a "dumping" ground for poor managers. In addition, the technologist ladder pay scale and titling were usually compromised in order to maintain the desirability of pursuing a management career. Good technologists became mediocre managers in poorly administered "DUAL LADDER" companies, because the career path for managers appeared more attractive.. There was no technical ladder within P&G, where the primary Corporate external image was on fast-moving advancement among marketing managers. A separate reward system was viewed by the Corporate Human Resource organization as inappropriate for a single function. It had to be available to all functions.

Bill James took the lead with a small group to study the subject.

> *(James) There had been previous studies of a possible dual ladder system within P&G, but were rejected as not really needed, and not desirable given the many different functional skills needed within P&G beyond just research skills.*

I had become the head of the Industrial Research Institute (IRI) Human Resource Committee. This gave me access to other company's approaches to recognizing technologists. I was particularly attracted to the 3M Carleton Society. It recognized, in a very public way, 3M's most outstanding technologists. There was an annual ceremony inducting new members and pictures of Society members were shown in a very visible central location. The only feature I didn't like was the Society also included some managers. I was concerned that a similar technologist society could be used for poor managers, and could reduce the importance of a focused career as a technologist.

I decided to discuss the "society" concept with Gordon. So I set up a meeting at the General Office, and tried to come "loaded for bear". The conversation went something like this.

> *Bill -" I have been impressed with 3M's approach to recognizing their top technologists. They have a separate company Society for recognizing individuals for outstanding innovation".*

I then reviewed in detail my understanding of how it worked.

> *Bill: "We can call it something like... the Gib Pleasant Society (after a former R&D VP).*

Gordon looked at me like I had 3 heads. "James you have a lot of good ideas. And then you have an occasional clunker. This is one of the clunkers. Back to the drawing board"

This was a major disappointment for me, and I was mad.

"Okay Gordon, we'll eventually come up with something that we all like and works here."

Several months later I was in the office at 7:30 - my normal time to get any discussions out of the way with Gordon before the raft of endless meetings . Phone rings.

> *Gordon: "James... I got it!" With great excitement.*
>
> *Bill: "What do you have?"*
>
> *Gordon: "The name for a technical society we are going to create"*
>
> *Bill: "Okay, what's the name?"*
>
> *Gordon: "We are going to call it, The Victor Mills Society!"*
>
> *Bill: "Brilliant, I think that would really work!"......so, although frustrated by the delay, I was clearly elated..*

Gordon had come around to understanding the need and value, and landed on the individual's name who clearly represented the level of achievement desired. It was named after one of the most prolific inventors and innovators in the Company. We had evolved to the concept of the VICTOR MILLS SOCIETY within the company - one into

which only the most accomplished and productive researchers would be inducted. They had to have products in the market that served as examples of their innovative skills.

(Brunner) At the time I was working for John Smale, the CEO. We clearly needed his full support to establish this program. The heart of the message to John was: " The thing top technologists value most is being recognized by their peers. You can give them lots of things,.. they're all nice, but when they're recognized in their peer group, there's nothing more rewarding."

John always believed that if a company is upfront on what it values, it likely would get more of it, and he urged us to get it underway.

The inaugural class of Victor Mills Society members was inducted less than a year later. Shortly after John's retirement, with Ed Artzt, the new CEO, headlining the award ceremony we introduced the first 12 Victor Mills Society Member. It was held at the convention center in Cincinnati, in front of the whole global organization, whether they were there in person or on telecast around the globe. Victor Mills, into his mid-90's, but proud and vibrant, was present.

Recipients arrived with their entire families and were presented gold medals on stage by Ed Artzt as their accomplishments were recited. The excitement and resounding applause were electrifying.

 It was the biggest deal we had done to recognize individual employees in the history of the company. People saw vividly that the company REALLY did care about people who did big things. It was truly a fantastic inaugural event.

Paul Trokhan, who received his award for the breakthrough "Structured Tissue & Towel" technology which propelled the Company's paper products, when asked, said " It was the absolute highlight of my and all of our careers at P&G. Further, it clearly established the expectation that technology is the foundation for the Company's future success. At the ceremony, CEO Ed Artzt said, " This is establishing our Technologist Hall Of Fame ... like baseball, this is our All-Star Team.....but the great news is our players are still actively inventing!"

# Innovative Technology
# By Innovative Technologists!

## The Charter Members of The Victor Mills Society...
## Honoring Excellence in Technology at Procter & Gamble.

**Raymond E. Bolich, Jr.**
BS, Chemical Engineering
Beauty Care Division
Cincinnati, Ohio

**Kenneth B. Buell**
MS, Mechanical Engineering
Paper Products Division
Cincinnati, Ohio

**Marion David Francis**
PhD, Biochemistry
Norwich Division
Norwich, New York

**Eugene P. Gosselink**
PhD, Organic Chemistry
Fabric & Hard Surface
Technology Division
Cincinnati, Ohio

**Frederick E. Hardy**
PhD, Chemistry
Laundry Detergent Products–
Europe
Newcastle, United Kingdom

**William D. McKinney**
BS, Chemical Engineering
Cellulose & Specialties Division
Memphis, Tennessee

**John J. Parran, Jr.**
MS, Chemical Engineering
Health & Personal Care
Technology Division
Cincinnati, Ohio

**Pedro A. Rodriguez**
PhD, Analytical Chemistry
Corporate Research Division
Cincinnati, Ohio

**Paul Seiden**
BS, Chemical Engineering
Food & Beverage
Technology Division
Cincinnati, Ohio

**Geoffrey Stanley**
MS, Chemistry
Bar Soap & Household
Cleaning Products–Europe
Newcastle, United Kingdom

**Paul D. Trokhan**
MS, Chemical Engineering
Paper Products Division
Cincinnati, Ohio

**Horst G. P. Wienecke**
PhD, Pharmaceutical Chemistry
Health & Beauty Care–Europe
Gross Gerau, Germany

The in-Company recognition was followed with a 2-page special discussion of P&G's new recognition system in Scientific American, replete with photos of each VMS Member and a discussion of the Society.

It became clear very quickly that we had instituted a very important motivating and appreciated reward structure in the global R&D organization.

Interestingly, the Victor Mills society also received wide acclaim throughout the company, as opposed to cries of favoritism. It led to various other functions (Marketing, Product Supply, Finance) establishing their own versions of outstanding employee achievements.

The formation of the society, however, soon registered strong input from all levels of R&D researchers on "How do I achieve a VMS award when there is no career path acknowledging advancement?" The organization was registering that we needed to go further.

We needed to develop recognition at other levels of achievement along with salary curves for the entire non-manager/technologist group of employees. It had to be adaptable for the disparate business units ranging from the highly mechanical-oriented skills of the paper organizations to the heavily 4-year college minimum for the health care entry-level employees. From the heavy chemistry training of the cleaning products organization to the unique sensory needs of the beauty care organization. And any new system had to be cost neutral to the current situation.

Bill James led a global team to tackle the challenge. External bench-marking was done, although it was clear that no existing system fit the needs. The major challenge of developing this new system for P&G took 2 years to develop. The original career progression system that evolved is shown below:

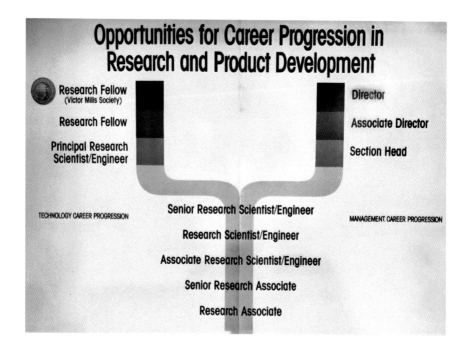

**Opportunities for Career Progression in Research and Product Development**

Research Fellow (Victor Mills Society)

Research Fellow

Principal Research Scientist/Engineer

Director

Associate Director

Section Head

TECHNOLOGY CAREER PROGRESSION

MANAGEMENT CAREER PROGRESSION

Senior Research Scientist/Engineer

Research Scientist/Engineer

Associate Research Scientist/Engineer

Senior Research Associate

Research Associate

The resulting career path plan was straightforward and simple. **ALL** R&D employees from their start in the Company could progress 4 levels to Senior Scientist/Engineer. **Technology Managers** could then progress three levels to Director, which was unchanged from the past. **Technologists** now could progress three distinct levels to the Victor Mills Society. This new career path plan was extremely well received by the Technologist community.

Overall, we were elated with the success of the formation of the Victor Mills Society, and the new career path plan for technologists. It provided great career clarity, as well as high motivation to our critical technologist community. It brought home the absolutely fundamental importance of a motivating reward system for a R&D organization.

Going forward, we tried to strongly underline for future stewards of the program the key potential danger to the Society's value to the Company. **That danger is the possible ongoing temptation to dilute the standards for achievement of the award**. This was seen as THE common downfall to other reward programs studied, and is of **PIVOTA**L importance.

# 12.

# KEEPING RESEARCHER SKILLS SHARP & ADVANCING

When a newly hired scientist or engineer joins the Company as a product development researcher, their interests transition from their years of formal academic training to the new skills required to be an effective member of a product development team. In a short time, researchers are facing real problems that need to be resolved. The new hire is soon focused on those team activities of most interest and where their skills are best applied.

There's a great desire, and value, in staying fully up-to-date on all of the latest learning that's occurring in the researcher's area of interest. This is a formidable task. In my experience technologic updates originated from the variety of product categories and technical centers in the Company. Further, the different global technical centers had very different histories, with spans of existence from 1 to 50+ years. There was a high need, therefore, not only to collect and transfer explicit learning…..but also to teach and train.

To facilitate this compilation and transfer, we established over time a network of 20 different "**COMMUNITIES OF PRACTICE - COP**", which at the beginning of the 2000 decade are listed in the TABLE. Of course, the areas of focus evolve over time as technology needs change.

## Communities of Practice

- Analytical
- Biotechnology/Life Sciences
- Colloid & Surface Chemistry
- 3D Visual Computing
- Fiber-Fabric
- Functional Polymers
- Imaging
- Malodor Control
- Microbiology
- Organic Chemistry
- Packaging
- Perfume/Flavor
- Process Community
- Products Research
- Regulatory & Clinical
- Robotics
- Sensors
- Skin Science
- Statistics/Math
- Wipes/Substrate-Based Products

Some communities have existed for over 40 years....some had recently been formed. The Analytical Chemistry COP is the oldest dating back to the 1970's. Each COP was sponsored by an R&D vice president, with a budget, and a mission to promote cross-fertilization and diffusion of knowledge. With the breadth of experiences that exist in a large global company, the value of this activity is critical. There's almost nothing that has not happened in P&G's long history, and if the positives can be captured and repeated, and the failures captured and avoided, the benefits on effectiveness can be a major competitive advantage. So each COP had the charge to convey both the explicit know-how published in reports and presented in poster sessions, as well as the tacit knowledge in researchers heads distributed by personal interactions.

In addition, the Communities of Practice served as a vehicle for researchers to gain peer recognition for their work through publications and presentations in seminars, and also for formal awards to be made for outstanding "community of practice" work.

Researchers could change which communities they want to connect with, and could change or add as they wish.

With the availability of the internet, there clearly was a great opportunity to facilitate communications and learning across the entire global Innovation community with an innovation website. **TECHNET** was established. The concept was that TECHNET could serve as a global "lunch table", just as an actual lunch table had served technology transfer for researchers when the Company was in its early years. Global researchers would all be in ONE building!

 After I retired, TECHNET continued to evolve. Nabil Sakkab, P&G R&D senior vice-president, reported on specifics after a number of years of operation*. "There's a mammoth target audience of 18,000 innovators across R&D, Engineering, Market Research, Purchasing, and Patent Operations. Sixty percent use TECHNET several times a week or more. There are 600 websites for Global Project Teams, and individual problem-solving and connection-making websites

for our 20 Communities ofPractice. This adds up to nearly 9 million documents on line, growing daily. Each and every month, over one million pages are accessed.

One of the most popular sites on INNOVATION NET was the use of SMART LEARNING REPORTS. Global researchers are required to capture key learning, and business-building insights from their experimental work via electronic reports, typically posted once per month. Every staff member could "mine" the insights from these reports via standard search engines, and text analysis software. The flow of information happens 24 hours a day, fetching knowledge based upon individual researchers' interests, identifying "virtual neighbors" with similar interests, and daily e-mailing links to reports on topics they have subscribed to. It was a marvelous capability that allows researchers to work in a global community that feels as if everyone is working in the same building."

Obviously the P&G system represented a very large-scale operation, but the principles and fundamental value of this innovation connections system would be applicable to an R&D operation of any size. It had tremendous capacity to accelerate company innovation…

....a true global lunch table!

*Research-Technology Management, March/April 2002

# SECTION THREE

## "THE MARRIAGE OF WHAT'S NEEDED....... WITH WHAT'S POSSIBLE"

Innovation is all the activities....the creative ideas that lead to inventions,,,,,that provide the technological basis for important product, package, process and equipment advances. You could call this the "WHAT'S POSSIBLE" part of innovation. Of course, this is all driven by the consumer. If there isn't consumer interest, there's no innovation. Just an interesting technical advancement. Consumers provide the "WHAT'S NEEDED" part.

Innovation involves the marriage of "WHAT'S NEEDED with "WHAT'S POSSIBLE".

### III

13. "THE CONSUMER IS BOSS"

14. "HOW KNOW IF SOMETHING IS WORTHWHILE?"

15. THERE IS NO COMMODITY.... JUST AN INNOVATION NEEDED

16. TIMING OF INNOVATION CAN REQUIRE PATIENCE

17. GET THE PRODUCT'S MARKET FOCUS RIGHT... (OLESTRA CASE STUDY)

18. "SERENDIPITY" - GETTING TO AN INNOVATION UNEXPECTEDLY

# 13.

## "THE CONSUMER IS BOSS"

Innovation is all about making things that people want to buy. It's not about the number of patents or new technical developments. It's first and foremost about the products and services you offer, and the resulting level of appreciation by end users. It doesn't matter if you market to consumers, to industrial customers, or as a device manufacturer. It's the ultimate product user who dictates whether you are successful.

Integrating and meeting customers' needs is the main task of a highly effective innovating organization. My shorthand for this is

> # "THE MARRIAGE OF WHAT'S NEEDED WITH WHAT'S POSSIBLE"

**"THE MARRIAGE OF WHATS NEEDED WITH WHAT'S POSSIBLE."**

It's about identifying the consumer or customer need, and coming up with novel and ideally protectable technology to meet that need. P&G focused early on in understanding consumer needs, pioneering door-to-door consumer interviewing in **1924.**

But consumers can't articulate their desires very well, particularly for radical innovations. So, you have to ensure that there's as much emphasis on the Discovery part of the innovation process as developing and commercializing new and improved technologies. It is often difficult to articulate the importance and skill required to obtain consumer understanding. Companies take the process for granted. The quality of in-depth consumer understanding and discoveries can make all the difference in achieving a great product result.

Take BOUNTY disposable towels as an example. Consumers said they wanted a stronger disposable towel. Stronger products were made and tested and consumers yawned. Then a key insight was uncovered. **"So strong that it won't tear when wiping a spill on a rug."** A different kind of strong! The focus was changed, and a **major** advance was created.

Another example was on LIQUID DETERGENTS. It was readily apparent from the unsightly drips over the outside of used detergent bottles in the laundry room that improved dispensing was needed. Yet when consumers were asked about this problem they responded, "I do need to be more careful when I measure." The need for a better measuring and dispensing device just didn't get expressed. It's that insight that led to the development of a self-draining cap on liquid Tide, which turned out to be a **MAJOR** breakthrough in consumer acceptance.

A third example was in the SHAMPOO area. Consumers kept feeding back to us, from product testing, that they wanted more lather. Products producing significantly more lather were formulated and tested, but they didn't generate the positive consumer response that was expected. Perplexed, the development group decided that they had to dig deeply to understand this conundrum. They installed a shower facility in the laboratories and brought in consumers who used the showers - in bathing suits of course - to test product prototypes. The users' comments were monitored during the shampooing process. It became obvious that when consumers talked about more lather they weren't talking about the amount of lather, but the **feel of the lather.** To address this, products were made with the kind of super creamy feel that consumers were looking for, and major advances in consumer acceptance were achieved.

I gained some insight into the importance of understanding the consumer early in my life when I was a golf caddy in the Chicago area. I was caddying at Rolling Green Country Club in Arlington Heights, IL. A member of the club was Ray Kroc, the founder of McDonald's. This was 1955, when Ray was working to conceptualize this great business. He was not the greatest golfer - but he was a great tipper. He loved interactions with the young caddies to find out their reaction to the products that he was serving in the first McDonald's he built in Des Plaines, Illinois - my home town. He would get passionate with us around our thoughts on the malted milks, hamburgers, and french fries - every detail on what was good, and what could be improved.. in painful detail! Further, he talked about another consumer strategy he had for his restaurants. In the mid-1950's if you were traveling on the road, and wanted a rest stop, you would be exposed to the worst restrooms possible! Dirty, broken fixtures, usually no personal care products….just awful for you and your family. Ray Kroc wanted his restaurants to be **THE** place to stop - with restrooms properly maintained. Customers may not initially expect it, but they would be absolutely impressed and grateful, and look to return in the future. What a perfect understanding of dealing with a consumer need, changing the game, and gaining a loyal customer. Ray Kroc drove home to me how much **passion** he was putting into the design of his products, as well as his McDonald's restaurant, to maximize consumer appeal. I sensed how important it was. **It stuck with me** as I moved on in my career, and it's a story I still enjoy telling.

While gaining insight into important consumer needs is normally tough, hard work, there is an occasional person with a real instinctive knack. In this regard, I have to mention my late younger brother, Vernon Brunner. He also spent his 40 year career with a single company - Walgreen's Drug Stores.

 He was Executive Vice President in charge of Merchandising. He brought the computer to pharmacy operations, drive-through and on-line shopping, and standard store layouts across all the Walgreen stores, along with great Company growth. Further, he had a great sense for judging the success of new products being launched, was always on the lead with winners, and Walgreens was never out of stock on major new successful product innovations. His completely unique contribution was an innate knack of knowing what customers would buy. He continually was first with very novel products that were immediate big hits - hula hoops, disposable cameras, chia plant sculptures, "silly putty" novelties, inflatable outdoor Christmas decorations, and many others. His acumen and successes were known throughout the industry, and at the time he was one of only two retail people inducted into the US Marketing Hall of Fame - along with Sam Walton.

He always focused on what was to happen. He said, "Knowing the past is fine, but to really succeed, you have to be able to look ahead." He was uniquely good at it for his customers. In today's world, he probably has been replaced by the product-analyzing Bot on TIK-TOK.

But, unfortunately, history is brimming with examples where this "instinctive feel" just doesn't work, and is a rare exception.

The importance of understanding the consumer is so strong at Procter & Gamble that responsibility for the consumer interface activity is shared between Product Development and Advertising. Product Development maintains a special group within its team called Products Research. This group of scientists and engineers is continually involved in consumer understanding, consumer testing, and generating insights that they can put together to help make better products. The value of this activity within the product team is critical because researchers always have a better idea of what's possible as they're viewing consumers' reflections on needs. They can be more open to radical ideas. Further, the research group usually has a much longer tenure in their jobs than the marketing people, who tend to move more quickly into other assignments. So there's greater continuity and institutional knowledge as the team progresses to create superior products.

The structure allows the identification of opportunities to be driven not only by what is learned from the consumer, or not only by a product technology, but by the effective integration of the two, leading to a stretching and consumer-preferred product vision. Some ideas get driven by a unique technology (i.e. FEBREZE), and others by an extremely obvious consumer need (i.e. PERT PLUS Shampoo & Conditioner in-one), but both have to come into play for success.

 Carol Berning was a researcher focused on the consumer in the laundry products area. She developed many important consumer insights which helped drive product technology work. Carol was able to document that a large percentage of consumers used smell before and after washing as most important to judge cleaning. That opened research to study specific odor profiles after the

wash cycle, and not just in the laundry detergent, and led to preferred fragrance designs.

Another important finding was that many consumers discarded clothing when it "looked tired", as opposed to "old, or out of style". Studying "looked tired" led to the development and use of a new cellulase enzyme which clipped the cotton pills that develop on the surface over time to restore the color. It led to "Keeps your clothes looking newer longer" Her long history of discovering important consumer insights led to her appointment to the P&G Victor Mills Honorary Technologist Society.

Once a key consumer need is in sight…the need has to be converted into a TEST METHOD. A reproducible test that can measure progress against meeting the need with a given product execution. This is a critical but unfortunately difficult step. Critical, because when you have a test that is focused exactly on resolving the consumer need, you can usually make very good progress. Difficult, because it is just not easy to take a broad based need - say, "better cleaning of my laundry" - and narrow it down to a specific test that will be representative of the consumer's decision-making on a given product's performance.

"Cleaning of laundry" might include removal of dozens of different type stains: garden soil, body soil, blood, whitening, and a myriad of other kinds of fabric soiling. While all are obviously important, what do most consumers focus most on when judging the performance of a detergent? That's a key question for designing a winning product.

The answer to this specific question could have changed at different historical times, as the makeup of laundry changed with the evolution of different consumer habits, household size, clothing styles and fabrics. But, at one point, and

perhaps still the case, researchers found that the primary test most consumers used for judging a given laundry detergents' performance was the clean-up of a "dingy white" item they had. That led to the global round-up of hundreds of global consumer"s "dingy white" items…extensive experimentation, new technologies, and yes,…..a consumer-preferred laundry product!

**The importance of identifying key test methods cannot be overly emphasized**. A faulty test method can lead to very frustrating failure, and the need for a complete re-do. But an on-target test method can be the biggest expediter of a highly successful innovation,

# 14.

# HOW DO YOU KNOW YOU'RE ONTO SOMETHING REALLY WORTHWHILE?

**I would always know by applying**

## the "WOW" test.

That is, when I describe the product and how it performs…and you respond enthusiastically…."WOW", I want to try that. Then when I was able to offer you the product, and you tried it and responded, "WOW" I want to use it again. **I can say that was the basis for a successful product. It never failed me.**

When you don't get these "WOW" responses, and you try to rationalize why not…..Look out! You're in trouble!

There's a popular story on the "WOW" test that was behind the innovation of the Mac computer, which really formed the explosive creation of today's global personal computer market. Steve Jobs was forming the team to create his breakthrough vision. He decided to take a prototype of the Mac, and cover it with a cloth. He would bring a promising candidate into the room, and dramatically unveil it. He would watch for the eyes to light up, the excitement to show, but most of all…..he wanted to hear "WOW". That was the critical clincher for identifying the passionate alignment with the product he was looking for, and for making the offer, and it didn't let him down.

It all sounds simple….right? We all know that isn't the case. It's difficult to really get at and identify a major unmet consumer need. Then it's equally difficult to develop a technical route to design and execute a practical product. And we haven't even touched on the challenges of scaling up from the lab prototype to a production version. However, when we were able to design a product that REALLY delighted consumers, we were almost **ALWAYS** able to find a way to successfully commercialize it in the marketplace.

**If you were to ask me if there is a "go-to technology" that was always the most effective and efficient at delivering improved consumer acceptance, I would say FRAGRANCE.**

Delivering the best fragrance that supports the key benefits of the product, during and after use, is critical to maximizing a positive response from the consumer. The right fragrance creation can support a wide variety of end-benefits …clean, fresh, pure, mild, strong, natural, healthy, efficacious, and a myriad of other sensations and images appropriate to a given product. Further, a specific fragrance like "APRIL FRESH" DOWNY fabric softener became a hallmark benefit for the brand… for decades.

P&G was the largest customer of fragrance materials in the world! We worked with every global fragrance supplier, who - at our request - would put forth the latest version of their most creative and sophisticated perfumes. These creative fragrances were expected to express desirable sensory attributes that supported the key benefits of the product.

P&G had internal world-class perfumers, who spent their entire careers at P&G, and became recognized for their expertise. These perfumers were assigned to a development team, and created the fragrance for the product, along side the team optimizing the product's functional performance.

Goeffrey Stanley was a self-taught perfume chemist, located in our Newcastle England Technical Center. He spent his entire career creating superior brand fragrance combinations, primarily on laundry and cleaning products - including "APRIL FRESH DOWNY". He was recognized internally as the most outstanding fragrance designer in Company history. He was one of the first group of the Victor Mills Honorary Technologist Society members in 1992.

# 15.

# "THERE IS NO COMMODITY.... JUST AN INNOVATION NEEDED"

A leading technologist never accepts that a product can't be improved in a major way. A classic example that I witnessed was the innovation launching JIF peanut butter.

P&G, given its experience with peanut oil as part of their involvement in vegetable oil processing, purchased the W.T.Young Foods Company in 1955. They were the producers of a small peanut butter brand, BIG TOP.

In the US, peanut butter formulation is controlled under a US. FDA "Standard of Identity". The product had to contain at least 90% peanuts, and **nothing else** other than salt, sweeteners, and hydrogenated vegetable oils.

From P&G's vegetable oil experience, a proprietary canola oil emulsifier had been identified to prevent peanut butter oil separation prior to the W.T. Young acquisition. It was invented by my first department head - Judson Sanders. He was an "engineer's engineer", who demanded clean labs. "A messy lab means a messy mind and messy research." Not surprisingly, we got the message, and stepped up to the task!

A touch of honey and molasses was competitively unique, added to Jud's formula, and this provided a small start to gaining a superior product. A new brand, JIF, was introduced in 1958.

At the time, SKIPPY had more than a 50% share, PETER PAN owned a mid-20% share, and JIF was just launching. Given the major constraints on peanut butter formulation, it looked like any "game changer" opportunity on JIF was **impossible**.

I was part of a small team formed to create a breakthrough on the JIF product. Our team was young, energetic, and highly motivated by the thought of **impossible.**

At the time, given P&G was involved in producing soap flakes, we had an old high-pressure soap flaking mill in our pilot plant. It had multiple roll flexibility to permit extremely fine crushing of materials passing through it.

Given we were open to trying anything, we put roasted peanuts through the mill, and gave it the full pressure treatment. Literally, the peanut cells were all fully opened. The resulting aroma was delightfully intense, and the flavor at a new higher than ever level.

The resulting peanut butter was smoother, and with an exciting new level of flavor and aroma.

But the excitement soon disappeared. With the exposure of the new desirable flavor and odor components, the peanut butter rapidly oxidized and the product quickly turned rancid.

The solution was radical at the time. The idea was to remove oxygen from the product, and replace it with completely stable and non-interactive nitrogen gas. Although common today for food packaging, it was entirely new for the JIF product.

The needed process was accomplished, and a new game-changing consumer-preferred JIF product rolled to the market.

Behind great marketing behind "CHOOSY MOTHERS CHOSE JIF", along with "TASTES MORE LIKE FRESH ROASTED PEANUTS, SMELLS MORE LIKE FRESH ROASTED PEANUTS" - the JIF brand roared ahead, and became the #1 brand in 1983. And it continued to grow share every year that followed. By the time JIF was sold to the J.M.Smucker Company in 2001, JIF  had a greater than 50% share, SKIPPY, which had been the leader at JIF's launch was around 20%, and PETER PAN was essentially out of business.

Certainly a great example of product innovation changing the game, and a "COMMODITY" basically yielding to determined attack.

# 16.

## THE TIMING OF THE INNOVATION MARRIAGE CAN REQUIRE PATIENCE

Bringing together "WHAT'S NEEDED" with "WHAT'S POSSIBLE" can require extreme patience regarding the right timing for that connection.

On occasion the technology for fulfilling a need is well ahead of consumer interest. I had such an experience in my first project at P&G . I was assigned to DUNCAN HINES MIXES, and was on a team to develop the first-ever **READY-TO-SPREAD FROSTING** product. Great tasting and easily spreadable formulas were developed. Stabilizing the formulas to avoid separation, oxidation, and spoilage was the challenge. Excellent frostings in a can were developed with the application of several proprietary technologies. This involved a butter-like vegetable oil blend, an advanced emulsifier system and novel mold prevention. It was a terrific product and the team was excited to see it move quickly to a test market.

Product

P&G entered a test market with a flair…but unfortunately the product didn't develop interest from the consumer. We had a new, patented first-of-its-kind product that bombed. Big time. Consumers in general couldn't see themselves moving away from their favorite "homemade" recipes to something in a can. P&G had a "once-burned, once-learned" memory on Ready-to-Spread frostings...and the idea was "deep-sixed".

This was the 1960's and women hadn't entered the workforce in large numbers yet. And the family size was still pretty large. A decade later huge demographic shifts had occurred.

It was **FIFTEEN** years later that Betty Crocker, and then Pillsbury, introduced spreadable frostings in a can to their bakery products line up. The timing then was just right. Women were ready to reduce some of the kitchen drudgery. The idea soon became the standard way that consumers tackled their frosting preparations. The technologies competitors used mimicked our earlier approach. P&G had entries on the shelf ready to go. But being first ironically resulted in P&G being last to enter the market. The technology was ahead of consumer interest, which later caught up. Overall, an example of really bad timing.

*P&G wasn't the only company who suffered the indignity of "too early introduction". Apple's famous Newton Message Pad is a case study in product failure. Introduced in 1993 when John Scully headed Apple, the device had a calendar, was capable of storing contacts and could even convert stylus scribbles into text...though poorly. It fit in a jacket pocket or purse and sold for $700. It was a MAJOR failure having no significant appeal. It was the butt of jokes on Doonesbury.*

*Ten years later the "Research In Motion" Company introduced the Blackberry with many of the same features. Its ability to replace faxes and communicate messages made it a huge success among business people, peaking with 85 million users in 2010.*

*Blackberry may have been the first connection device that worked in this area, but of course Steve Jobs and Apple, shortly brought forth the disruptive I-PHONE. That redefined the portable phone and personal digital assistant for ALL consumers, and Apple followed with new innovation advances for these products continually for the last 15 years.*

. . . . . . . . .

More often, product technology creation lags, and never seems to satisfy consumer interest.

A classic example within P&G was improving convenience in laundry detergents, and the evolution of **CONVENIENCE** and also **SINGLE-DOSE** detergent products.

Consumers historically appeared very interested in more-convenient laundry detergents. There was a natural evolution with P&G in the lead…most of the time.

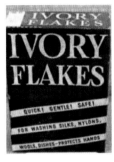

Ivory Soap Flakes was launched in 1923. Soap flakes took the place of soap bars for doing laundry, and provided a significant increase in convenience. When the major discon-tinuity of granular deter-gents later emerged as a potential reality, Richard

Deupree, the P&G CEO said, "**If anyone is going to obsolete our laundry soap, let it be us**".

Granular detergents then became the preferred form with TIDE in 1946, as they provided a major advance, not only in cleaning, but in major dissolv-ing convenience.

Pre-formed single-use products emerged when P&G introduced a new compressed detergent disc in the early1960's called **SALVO**. It was heavily marketed as the NEW FORM for detergents, but it had major issues dissolving quickly enough on the one hand, and frac-turing before use on the other. Further, a common comment - "I love to be able to use it over and over again".. Great effort was put into solving the negatives, but it was finally discontinued in 1974.

A unit-dose sheet, called **TOB JOB ,** which provided a revolutionary way to combine detergent with a high performing oxygen bleach was launched in Germany in the late 1970's. It was very well received by German consumers interested in maximum cleaning performance.

However, it got attacked by strong German environmental organizations for "excessive" focus on laundry problems. Whaaaat?

A similar unit-dose sheet, called **TIDE MULTI-ACTION SHEETS** was test marketed in the US in the mid-1980's. It combined detergent, oxygen bleach and fabric softener in one. It was an engineering marvel, using non-woven fabric pouch separation in the sheet to avoid interaction and stability of the detergent and bleach until the wash cycle. A softener coating on the sheet then activated in the clothes dryer. Although it provided a new high level of convenience and cleaning performance for US consumers, the product was not expanded. It certainly was a vivid testament to P&G's determination to try to meet the consumer need for laundry convenience.

Automatic dishwashing detergents in Europe started moving to **compressed tablets** in the late 1980's, importantly supported by European automatic dishwashing machine manufacturers. The tablets quite rapidly gained consumer acceptance over powders for their convenience. P&G's product creatively separated enzymes in a surface addition to the tablet that were desirable to dissolve quickly at the lower pH in the cycle from the higher alkaline ingredients.

However, the tablet form had significant disadvantages:

- The compaction force was intense, and caused issues with interaction of key ingredients, like bleach and enzymes, as well as dissolution issues in the dishwashing process.

- Key liquid ingredients in the formula had to be limited in order to maintain tablet integrity.

- Consumers disliked direct skin contact with the tablet for safety reasons.

At the time, technical advances with PVA (poly-vinyl alcohol) films offered the possibility of a pouch or pod to be formed which was durable, shippable and storable, but would dissolve when exposed to water. Further, high auto-dishwasher water temperature eliminated a dissolution challenge.

The European-based Global Dish Care R&D team, led by Claude Mancel in Brussels, Belgium, worked with the US-based Monosol Company to develop the PVA specifications. They also engaged the US arm of the Italian company Fameccanica for the unique multi-component proprietary pod making process. All the shortcomings found with the original dish-washing tablet form were resolved with the **POD** innovation.

The pod executions permitted a combined powder/liquid pod in CASCADE and FAIRY auto-dishwashing products. Ingredient compatibility issues were solved by also providing multi-compartments in the liquid pod. The different liquid compartments were indicated with different colors to provide a very eye-catching feature.

This technology produced the global, auto-dishwashing pod and the base for moving to laundry detergent pods.

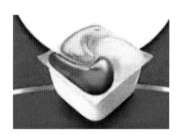

For detergents, additional development work was needed on the PVA film for cold water dissolution. Further, it was possible to go to an all-liquid pod in both TIDE and ARIEL laundry detergent pods, and the need for a powder to be included was eliminated.

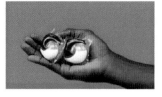

Thus, a new step in convenience was reached for detergent consumers with the **UNIT-DOSE PODS.** The pod form has been rolled out globally on all P&G detergent brands over the last 10 years, and has rapidly built popularity with consumers. Whether this latest form proves to be the final step in convenience for detergent consumers remains to be seen……..but, with 125 years of past innovation progress, I doubt it.

In fact , currently P&G is test marketing a new brand, called EC-30. It's a **dissolvable laundry sheet** - a single sheet for a laundry load.

The photo shows the market package and key selling message......
CARBON NEUTRAL CLEANING. Clearly designed for maximum environmental benefits and requiring major technology advances. Like any new technology or product, the consumers in the test market will provide the level of interest and learning for its evolution.

· · · · · · · · ·

For an extreme case of "WHY DID IT TAKE SO LONG", I think of the addition of wheels to suitcases.

This really didn't happen until 1970. Bernard Sadow is credited with the first patent, and his product was first sold in Macy's. Initially, the  wheeled-suitcase was considered appropriate and appealing only for fashionable women. Today, no-one would be without wheels for even the smallest traveling bag.

 Before wheels on a suitcase, travelers lugged, dragged, and portered luggage wherever they travelled…schlepping.through airports, train stations, down streets, and in and out of homes, offices, stores, and manufacturing plants. Various "luggage dollies" were invented and sold, which permitted luggage to be strapped on (if lucky). The better way of merely attaching wheels wasn't seen. **The "wheel guys" didn't connect with the "luggage guys".** It clearly illustrates how illusive an innovation connection can be.

# 17.

# GET THE PRODUCT'S MARKET FOCUS RIGHT

## "THE OLESTRA STORY"

One of P&G's greatest technical inventions was the discovery of the no-calorie fat - OLESTRA.

*The OLESTRA saga is a 28 year adventure in brilliant discovery, great chemistry, terrific engineering, strategic errors, and the power of vicious, faulty publicity. The adventure starts with a serendipitous discovery by Dr. Fred Matson, a brilliant organic chemist, in corporate R&D at Miami Valley Laboratories (MVL) in 1968. Researchers were synthesizing substitute fat molecules, looking for a fat that would be better ingested by premature babies. In the process they found a unique fat molecule that had the same taste and cooking properties of normal vegetable oil, but a molecular structure too large to move through the intestinal wall, and therefore it contributed no caloric or nutritional value.*

Normal fats have 3 fatty acids combined as an ester on a glycerol backbone. OLESTRA, by comparison, has 7 or 8 fatty acids attached as an ester on sucrose (sugar). It was a brilliant discovery.

Jim Letton, an outstanding chemist who later was inaugurated into the Victor Mills Honorary Technologist Society, did the unique tailoring of the OLESTRA molecule to balance eating and digestive properties.

The first application attempted was a submittal to the FDA as a "drug" to lower cholesterol, which OLESTRA did. Lengthy studies, however, came up short of FDA's minimum 15% reduction affect.

In a 1983 product development review, John Smale, concerned about the slow pace of OLESTRA development at MVL, ordered that the project be moved to the Food & Beverage Product Development organization. Bill James was the head of the organization at that time, and **Bill will continue the story.**

We identified 3 major hurdles that needed to be resolved.

1. There had to be an initial product entry that had strong consumer interest.

2. We had to develop a process to make the unique Olestra oil on an industrial scale - a very significant challenge.

3. We had to develop a regulatory strategy and necessary clinical support to achieve FDA safety approval,

Understand that this unique molecule could be used in just about any fat-containing product application. It made a fantastic margarine, potato chips and french fries were indistinguishable from the regular products. It made a variety of excellent frying fats including industrial applications for restaurants and other food preparation businesses.

The product work was led by Bill Hahn, a very experienced Products Research veteran. Bill laid out a product evolution tree that started with margarine and Crisco Oil - a prime choice because of the known heart health benefits. It then evolved to deep fat frying applications and finally to snack and bakery items like Duncan Hines cake mixes and Pringles potato chips.

To get the process development scale-up work started, I leaned on an old friend in P&G's Chemicals Division, which was P&G's internal commercial oil and chemical manufacturing division.. I engaged Dr. Ron Sampson a PhD chemical engineer and a great fat processing expert. He and I worked out the process steps to achieve feasible OLESTRA production on a large scale. Although glycerol esterification with unique fatty acids was a common process to produce custom oils, there was no large scale experience with sugar esterification. The physical properties of the combined materials during sugar esterification were more challenging, and different equipment design, reaction conditions, and transfer and storage requirements needed to be defined. The challenge to achieve OLESTRA commercial feasibility was of such great interest, that we were able to get the best engineering experience available in the Company to demand to get assigned to the project. A feasible process evolved, and later was built as a full-scale manufacturing plant in Cincinnati.

Back to the product. The margarine application had clearly positive aspects. There was the heart health benefit, and also P&G was now in the refrigerated business with CITRUS HILL orange juice. However, margarine was rejected. A major hurdle was that we weren't in the margarine business, and it would require great new learning. Further, if FDA classified the product as a drug, we might need a separate location in the grocery store. Further, the dairy case was controlled by Kraft, one of our competitors. Cancel margarine.

The next application attempt was as a food additive. We negotiated with the FDA on a position that OLESTRA could replace 35% of the fat in home cooking and 75% in commercial uses. Three years of rigorous testing was undertaken. FDA regulators then concluded, however, that because of its caloric properties there was a risk of "over-consuming" OLESTRA. The long-term affects of very high levels was not documented. It was not approved and long term extensive studies would be needed.

### Back to Gordon:

In 1990, 22 years from the discovery, the focus was narrowed to seek approval for "savory snacks" - potato & tortilla chips, crackers, and similar foods. This market focus was a PIVOTAL decision, and hotly debated internally in the company.

Clearly, " salty snacks" offered a huge market and financial opportunity. There was great excitement expressed by Frito-Lay, the leading snack producer. With them as a partner it provided a formidable duo with a dominant voice for the snack application.

But there also was a strong point of view to enter with a "health-oriented" line-up of foods. Pre-prepared foods specifically directed at providing doctors and health-agencies new ways to help patients with heart disease, obesity, diabetes, etc. The idea was to create a strong **health pedigree** for OLESTRA before trying to enter a market like

 One of the board members, Dr. Ralph Snyderman, who was Chancellor for Health Affairs and Dean of the School of Medicine at Duke University, was a very strong advocate for this entry route.

The fact that key patents were expiring in 1996 was another factor finally leading to the decision to proceed with "salty snacks" The company's extensive human safety studies were extremely positive, providing confidence in the product.

FDA approval came in 1996. The product "meets the safety standard for food additives, reasonable certainty of no harm". A labeling requirement was mandated…"Olestra may cause abdominal cramping and loose stools. Olestra inhibits absorption of some vitamins. Vitamins A,D, E, and K have been added."

 Olestra was branded as OLEAN, and launched as the frying oil in Frito-Lay's new brand WOW potato chips. The product **rocketed** off shelves.

Unfortunately, the product served as a welcomed platform for "The Center for Science in the Public Interest", founded by Dr. Michael Jackson. This organization was formed by seasoned lawyers from Ralph Nader's highly influential consumer advocate effort against automobile safety. Nader had become famous, and actually a four time US presidential candidate. Far from a science  organization, CSPI promoted consumer advocacy for safer/healthier food. In the Washington press, they were known as the "food police"

OLEAN provided a perfect, high-profile target for them, and they used every possible communication avenue to attack the brand. They were effective in raising high level concerns with consumers, and also were effective in drawing in large donations for their support.

In 2003, the FDA withdrew any requirement for the full mandated warning label, based on deep reviews of post-market studies submitted by P&G. The FDA concluded and publicly reported that subjects eating Olestra-containing chips were no more likely to report cramps, loose stools, or any other GI issues than subjects eating an equivalent amount of regular chips.

But…the damage had been done…irreversibly….and OLEAN snacks faded and eventually were phased out.

It certainly was the biggest product failure Bill & I were involved in. It was a true breakthrough discovery. There was a fantastic commitment by all P&G groups to commercialize it. The human safety testing work was exemplary. **Opportunity LOST. DEPRESSING!.**

There's obviously no way to know…..but, in retrospect, I've come to believe that the extremely bold entry plan behind a major new food item like snacks….was just too high of a risk. An alternate entry plan of a health-oriented food product, like a margarine with heart health benefits, or a line of pre-prepared "health-oriented" foods to gain medical experts' support might have worked. We'll never know.

Clearly, a PAINFUL underscoring of the critical importance of the product choice and market focus for a major new technology! But also a vivid example of P&G's commitment to commercialize what was clearly a technical breakthrough invention.

# 18.

## "SERENDIPITY" GETTING TO AN INNOVATION "UNEXPECTEDLY"

"We're at that awkward stage of trying to explain a completely accidental discovery as our intended objective."

In the world of innovation, there's always the unexpected situation when something goes terribly wrong, but there's also the situation when something goes terribly right.

I think back to two particular examples where fortunately, something went **very right.**

 In the early 1960s, Palmolive was leading the dishwashing liquid category on a superior mildness strategy. Most people can remember the commercial, which still runs today in many parts of the world with Madge, the manicurist character soaking hands in Palmolive liquid claiming "You're soaking in it. - Takes care of your dishes and your hands"

Procter and Gamble decided that they wanted to invest in a program that would develop a superior mildness surfactant for dishwashing liquids.

A top team was put together at our Miami Valley corporate research headquarters in Cincinnati, and terrific work was done. A superior surfactant was identified - a unique amine oxide. Clinical studies showed that it was clearly superior in mildness to the Palmolive product.

Furthermore, it was an entirely new surfactant, which led to a patent application and approval.

In the laboratory, tests were underway to learn about this product in all kinds of situations. One of the researchers, a very seasoned veteran, Joe Blinka,, did some experimenting with the product in heavy grease situations. He found that the product was HIGHLY effective in forming quick emulsions with oil and grease. He pushed hard with the marketing group that this was a breakthrough in grease handling, and urged that the product be positioned on this platform. There was great resistance to this, given that the market experience suggested that mildness to hands was really the most important benefit to consumers. But here there was a product that was very mild to hands, but had a completely unusual property in being far superior in cutting through grease, which of course was a common problem in dishwashing situations.

GETS TOUGH GREASE BETTER THAN THE LIQUID YOU USE NOW.

Based on Joe's urging, a concept and use test was run with the product focusing on grease cutting properties. Consumers LOVED the product. A test market with the brand name established as DAWN was launched, and proved to be an outstanding success. DAWN has built share in the dishwashing market ever since until today, and clearly is the number one dishwashing liquid in the United States, and in other parts of the world under the Fairy brand name. A serendipity occurrence, but one of the strongest brands that ever was created.

A crisis situation in 1989 with the sinking of the Exxon Valdez ship in Alaska, and the spilling of 11 million gallons of oil brought out the unique mild & tough-on-grease properties of DAWN. It proved invaluable in the cleaning and life saving of ducks affected by the disaster. Since that time, DAWN has developed a

unique image for its safety and effectiveness in dealing with animals in similar distress.

The other serendipity situation involved a personal experience that I had. I had just been transferred to Europe, and was getting settled in my new job as the bar soap and household cleaning products R&D director.

One of my first activities was to visit the general managers of the different European countries to introduce myself, but also to find out how I might be best able to meet their specific needs.

In my visit to Germany, I had a meeting with the general manager, Helmut Fischer. When I probed what his most significant problem was, he surprised me with his answer. He said, "Right now, we're having a terrible issue with a very small brand 'REI in der TUBE.'" This was a popular variation of the REI detergent brand, which was used by German consumers when traveling and needing to do laundry in a hand wash situation. Helmut went on to say, "The product has lost its stability, and we're seeing product separation all over the country and are bombarded with consumer complaints."

I told him I would take a look at it quickly. As I explored the situation with my people in Brussels, there was very little known about this product. It was acquired many years before from a German chemist, and was given little attention over the years. My people said that the complexity of the formula using very unusual surfactants and stabilizers seemed foreign to modern researchers.

I looked at the product and actually saw that it reminded me of a product that I was familiar with in the United States – PRELL Concentrate. PRELL Concentrate was a shampoo product sold in a gel form in a tube. It happened to be green.

I said to my people, let's get some PRELL Concentrate, see if we can formulate it with a white color, and see if it might be a quick replacement for the current REI in a Tube product. We did that. It looked great. We put it in a quick consumer test. It scored clearly superior to the existing REI product. It also happened to save 40% in cost. We quickly moved to transition "REI in der TUBE" to this new formula. It was launched, was loved by German consumers, and Helmut was elated. Another example of **lucky product development - "SERENDIPITY"**

When thinking about serendipity innovations, one has to remember Sam Walton, when he launched his first mega-store. The clearly logical place was a large city location with a huge base of potential customers. However, his wife, Helen, refused to locate in a large city. So, he chose Rogers, Arkansas, population 5,000, with loads of good quail hunting for Sam. Well, we all know it was a runaway success! Sam said: "It was the very first lesson we learned. There was much, much more business in small-town America than anybody, including me, had ever dreamed of. And no significant competition either"

# SECTION FOUR

## "THE PRODUCT INNOVATION MACHINE"

# 19.

## THE IMPORTANCE OF BUILDING FROM YOUR TECHNICAL STRENGTHS AND EMBRACING OPEN INNOVATION

A Company's technology assets are of tremendous value. They represent the heart of the company's products, and have benefited from competitive wars, years of improvements and refining, and people assets behind them with critical knowledge. Any company, as it grows and develops or acquires new products and technologies, has a portfolio of technology assets in its possession. The attached chart is a graphical depiction of P&G's brands evolution, and the range of technologies in the Company's arsenal, as it existed in the early 2000's.

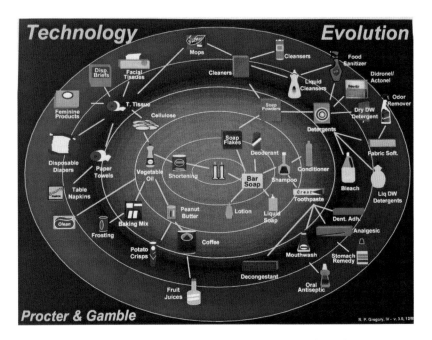

The very first observation from this graphic is that the Company expanded by building off its technical strengths, and by connecting different technologies to make new unique products. Connections have been in P&G's blood…its part of the Company's DNA and heritage. Along with acquisitions, it's how the Company successfully grew and evolved its business.

In the beginning, P&G created candles…and they were good. Candles begot soap from animal fat renderings. And animal fats led to a revolution in food fats with vegetable oils and CRISCO, the first all-vegetable shortening. Being strong in oil technology, P&G evolved into emulsifiers and surfactants, critical components of synthetic soap products - TIDE laundry detergent….then shampoo….and dishwashing liquid…and a variety of household cleaners…and this all was much better. The processing of vegetable oils, like the crushing of cotton-seeds, enhanced our understanding of cellulose, the technical basis for paper products, and then to feminine hygiene products, and sheet-delivery of ingredients like BOUNCE, and then to SWIFFER floor cleaning. Detergents taught P&G how to control calcium in hard water, which led to Crest, and calcium-fortified beverages, and on to controlling osteoporosis with ACTONEL…**And on it goes….extremely good!**

The learning here is that **building from your technical strengths is a VERY sound growth strategy.** P&G's, nor any company's technical base has been exhausted. There still exists ample opportunities for new discoveries that provide additional valuable product assets. Further, the creative combination of technical strengths can produce additional exciting products.

When a company like P&G was small, the technologies could be connected literally at the "lunch-table" by technologists from the different product areas. As the company grew in size, and globally dispersed, making those critical connections became a major challenge. An approach that was put in place to tackle that challenge was the "Global Technology Council", which is discussed in Chapter **29**.

As the Company grew even larger, decentralized, and global, connections became much more difficult. Further, one can't be solely internally focused…P&G has no lock on all the connections that can build the business. An increasing number won't be chemically or biochemically or paper based.

Important global innovations were increasingly being done at small and midsize entrepreneurial companies. Individuals were eager to license and sell their intellectual property. University and government labs had become more interested and aggressive in forming industry partnerships. They were hungry to monetize their research. It was estimated that for every P&G researcher, there were 200 scientists or engineers outside of P&G who were excellent technically - a total of perhaps 1.5 million people whose talent we could potentially use.

The internet was increasingly becoming a leading way to make connections on almost any subject. It was clear the Company needed to move to OPEN INNOVATION - no longer just internally focused. Creating new connections with whomever has the ability to help the innovation process, and bring major new product advances to the marketplace.

# RESEARCH & DEVELOPMENT HAD TO DEVELOP A SIGNIFICANT COMPONENT OF "CONNECT & DEVELOP" CAPABILITY

We decided to take a VERY BOLD step. A **one-of-a-kind** "deal-making technology trading expo" was planned and launched as "INNOVATION 2000" This major undertaking is discussed in Chapter **30.**

# 20.

# ESTABLISHING THE EXPECTATIONS

## SETTING THE COMPANY PRODUCT GOALS and A COMMON WAY TO MEASURE RESULTS

In every company it's important that the product goals and expectations are extremely clear. In a large company, like Procter & Gamble, it was challenging to set common expectations across the many product segments that existed.

When John Smale became CEO in 1981, he was unhappy with the Company's record in developing big new products. At that time he decided to transfer and promote Wahib Zaki, a very demanding leader from Europe, to assume responsibility as CTO reporting to him.

Wahib Nassif Zaki was a true "one-of-a-kind" individual. He was from Cairo, Egypt from a very prominent family, and was schooled in the finest schools in Egypt. As a youngster he was able to travel broadly with his family. He did university schooling in London, and graduated as a PhD in chemical engineering from Imperial College.

He joined P&G at the European Technical Center in Brussels, Belgium. He quickly made a highly notable impression with his technical talent, great energy, natural leadership, and direct and clear communication skills. He was a clear candidate for advancement, and was now in charge of the entire European R&D organization.

Zaki was world traveled, a bachelor, fluent in 3 languages, conversant in several others, and loved major European cities and their hotels, restaurants and entertainment. A real "bon vivant". He had worked in Cincinnati before (BOUNCE - (Chapter **22**) and coming back was very low in his personal desires… but with very strong urging from Smale..he came.

In replacing Harry Tecklenburg as CTO, Zaki was replacing a legend in the paper business, since he had been intimately involved with every aspect of tissue and towels, Pampers, feminine hygiene, and all the paper products' tremendous growth. Harry had a strong record of innovation, implementation, and major cost achievements in the paper business. Further, he had taken the lead to move the Company into Pharmaceuticals, had shepherded the recent acquisition of Norwich-Eaton Pharmaceuticals, and now was to head up that business.

Harry had been CTO for eleven years, and had developed a broad level of trust and comfort among R&D employees with his laid back, hands-off, low- interference style with operations. It was going to be a definite awakening of the organization with the arrival of the new hard-driving CTO boss, Wahib Zaki.

Zaki was charged with "changing the game." He quickly adopted a mantra called ;

## "Big Edge Product"

and made his expectations clear in any interaction he had. Further, he provided the following definition in a message that he sent to the organization :

> **"First of all, what do we mean by a big product edge? We all instinctively know, but we may express it in different ways. Rather than attempt a bulky definition, I will list several key characteristics. First, the product must deliver performance and acceptance benefits which are clearly superior to what exists. Second, the product must be presented expertly to the consumer and in a timely manner. Third, the product must provide a solid basis for volume and profit. A fourth factor invariably connected with big product edges is new and proprietary technology. Equally important to all of this is the recognition that the Big Edge is a moving target, and that the ultimate judge is the consumer in the competitive marketplace."**

With the challenge given to the entire Company R&D organization, CEO John Smale followed up with a letter to all the business units underlining what Zaki had said and strongly urging them to adopt this approach in their operations. Further, he would plan to engage them in reviewing their Big Edge Products and Processes in future annual Product Reviews.

"Big Product Edge" had to be measured. Procter & Gamble did a ginormous amount of consumer research, with experimentation in just about every different way to test consumers' reaction.

P&G did paired-comparison tests, single product tests, concept-aided single-product tests, brand identified single product tests and a myriad of other types. From all of this, a straightforward single-product blind test was chosen as the standard for measuring a given category's strength versus competition. Fundamentally, this was an approach where the product was put in generic, blind packaging, used at home and the consumer was then asked to rate it versus their expectations and normal product.

It was a very rigorous and highly challenging test. It took great product performance for the consumer to see a significant advantage versus the product they used and often loved! This comparison, across all of our business categories, gave us a very good helicopter-view of our product performance and product positioning versus competition.

John Smale reinforced the message by setting up annual product reviews with every business category across the company. Zaki accompanied Smale in each of these reviews. In this review an R&D director had to be prepared to put their product portfolio up for critique in front of the CEO and the CTO. This approach provided challenging and very clear objectives to all the R&D organizations and, of course, to all of the general managers in our business units.

When I eventually took Zaki's place as CTO, we continued the product reviews with the CEO. My position on our product position was always

# "If the business isn't good, the product isn't good enough."

I found this simple expression to be very, very accurate. Unfortunately, it probably caused a lot of indigestion among my R&D managers.

Later when I was on the Board of Directors, I instituted an annual product review with the Board across all of our brands and all of our business segments. I would show the situation where we had a clear competitive advantage… where we might be at parity…and, uncomfortably, product situations where we were deficient and inferior compared to the competitor.

It was amazing and frankly quite startling to the Board members to see how much correlation existed between the strength of the product and our market position. It certainly highlighted the importance of superior products to the Board of Directors. Just as we felt so strongly about it internally, this critical assessment of product superiority versus competition and business performance became a regular part of the annual Board product review.

# 21.

# GLOBAL CAPABILITIES ADD MAJOR ADVANTAGE

Procter & Gamble started internationally in Europe in 1930 in England and over the years created technical centers in Newcastle, England, Brussels, Belgium, Frankfort , Germany, Rome, Italy, and Paris, France. A center in Caracas, Venezuela supporting Latin America began in 1957, In Osaka , Japan in 1970 and in Beijing, China in 1999. At the Corporate headquarters in Cincinnati, Ohio, six technical centers were operational during my tenure…Additional centers entered the picture at different times from Company acquisitions.

## WHAT ARE THE ADVANTAGES OF THIS GLOBAL TECHNICAL STRUCTURE AND CAPABILITY?

### THE ABILITY TO ATTRACT THE TOP TECHNICAL TALENT

The most important factor behind the Company's success has been, will likely always be, the ability to attract the most highly capable people possible. Operating globally with on-the-ground capability to attract such technical talent, provides a major diverse talent advantage opportunity.

THE CRITICAL CAPABILITY TO UNDERSTAND THE REGION'S CONSUMER NEEDS, THE TECHNICAL STRENGTHS AND

STRATEGIES OF COMPETITIVE PRODUCTS, AND SUPPLIER
& UNIVERSITY CAPABILITIES

- **CONSUMERS**

  Researchers learn from close contact with consumers and
  exposure to new technologies that originate outside the US.

  There are so many critical examples where consumer input
  from one geography helped develop benefits for all categories.
  To be a global leader, it is essential to design products for the
  most demanding consumers in the world.

  The prime example is the Japanese consumer. In my experience
  there is no consumer more discerning, and more helpful in
  providing insight to perfect a product. They have an inner drive
  for perfection - in everything they do, and for the products they
  use. The features of the package - the ease of handling, storing,
  opening, using, and its appropriateness for the environment
  and responsible disposing. The detailed characteristics of the
  product.- color, odor, in-use features with nothing overlooked,
  and discovery of shortcomings with starkly candid descriptions. Whenever we thought we had everything worked out,
  the Japanese consumer would put the team back to work. And
  their input would drive benefits for the entire world.

  But there were many other examples : The high benchmark
  for laundry whiteness in Italy, the criticality of laundry product suds performance in European front-loading washing
  machines, as well as Latin American hand-washing. Cold water
  or used bath-water habit for laundry in Japan, severe hard
  water situations in various parts of Europe. Extensive stone
  floor cleaning challenges in different countries. Small package

design needs for countries needing small-size purchases to meet income constraints. The very different diaper-changing frequencies in different countries altering consumer satisfaction with different diaper designs. And, fragrance preferences vary markedly in different parts of the world......with high-impact fragrances very popular in Latin America...... and highly delicate fragrance critical in Japan.

Where we marketed without good consumer understanding, and thought what worked in the US would be fine, we made a lot of mistakes.

- **COMPETITIVE PRODUCTS and STRATEGIES**

Monitoring global competitor product innovation is, of course, absolutely critical. Again, Japan was always a major innovation threat. They were the first to introduce 'compact" detergents - high density granules allowing very small packages and reducing package sizes by 75%. Compact detergents were preferred by consumers everywhere, and became the standard for granular laundry detergent products.

Japanese competition was also the first to introduce ultra-thin diapers (the addition of chemical adsorbents into the cotton structure)......and pull-up training pants.....both of which challenged P&G in a major way to respond, and which again became standards for global baby products.

Japan competition were also first to bring initial implement concepts for easy dusting and floor cleaning, which drove our efforts on SWIFFER, where we were able to gain global market leadership.

European laundry product competition was always at the leading edge in oxygen bleach technology, and drove P&G to create the breakthrough bleach catalyst that was first introduced in WORLD LAUNDRY GRANULE (Chapter **24**) detergents in order to gain superior performance.

- **SUPPLIER AND UNIVERSITY CAPABILITIES**

Close relationships with key suppliers is a pivotal interface for any company, but critical for a company wanting to be a leading innovator. For an American-based global company, key suppliers are based in different parts of the world.

Interesting is that the development of proprietary surfactants and polyphosphates, which formed the basis for P&G's leading brand TIDE, was done in a joint research & development project with German chemical suppliers in the mid-1940's. So, the fundamental importance of these relationships is well understood.

A key competitive strength for P&G were the close technical relationships that have been developed with major suppliers in Europe - facilitated by the on-the-ground technical centers established there. These supplier relationships provided critical technical advantage for the entire global operation.

Some of the key strategic relationships were developed with the following innovative companies:

HOESCHT, BASF, EVONIK. - German leading chemical suppliers

NOVOZYMES - Norway enzyme innovator

INTERNATIONAL FLAVOR & FRAGRANCES (IFF) & FIRMENICH - Swiss global fragrance leaders

SOLVAY/RHODIA - Belgium & France polymer and chemical supplier

TETRAPACK - Swedish innovative packaging supplier

More of the value gained from these supplier partnerships will be covered in a later chapter on "OPEN INNOVATION" (Chapter 28)

Beyond the strong interfaces that can be developed with suppliers, there is also the opportunity for staying close to outstanding research at global universities. For example, P&G has had a long interest in studying bone and teeth function and disease, and the effective control of calcium. University researchers in Britain and Switzerland helped lead the way, and European clinical studies and medical scientists were very much involved in our pursuit of a more effective treatment for osteoporosis.

## RIGOROUS INTERNAL PEER CHALLENGE OF THE PRODUCT

Given the internal global testing in the different continents, the range of testing promotes uncovering any weaknesses or particular strengths in the product's performance. In laundry testing for example, the range of high European boilwash temperature of sophisticated machines to the cold river water hand washing in India provides true stress test experience. This yields important knowledge on every element of performance - solubility, sudsing, soil removal, water hardness differences, rinsing, etc. The data provides guidance for all geographies to deal with relevant shortcomings, and, of course, exploit particular strengths.

This kind of broad global testing provides great learning in every product category.

# 22.

# TECHNICAL EXCELLENCE IS AN ABSOLUTE MUST

The RESEARCH & DEVELOPMENT organization has to establish the highest standards for its technical work. Of course, hiring the very best technical talent is Job #1. Further, establishing the highest standards for experimentation, documentation, reporting, peer reviewing, and publishing the work…..requires employee training programs to infuse those standards.

All of this is critical for patent applications, claim support, regulatory approvals, and for the orderly transfer of technology and development made in the labs. It gets even more critical within global organizations, where critical findings need to be transferred for application and replication in many different geographical locations.

There were countless situations where the quality of P&G work was critical in gaining important external approval. Three specific examples come to mind which illustrate this point.

### 1. US MEDAL OF TECHNOLOGY

The US Medal of Technology was established in1980. It honors the outstanding individual scientists or companies in the United States for accomplishment in the innovation, development and commercialization of technology, as evidenced by the establishment of new, or significantly improved products, processes or services. It is administered by the Department of Commerce for the President of the United States.

It is the **highest award** the United States bestows for achievements in technology. It primarily focused on individual achievement, awarding great technology leaders like Steve Jobs, Bill Gates, Gordon Moore, Edwin Land, and W. Edwards Demming. Although only five Companies had received the award in the previous 14 years, Procter & Gamble Company received the award in 1995. P&G was the first consumer goods company to ever receive this major innovation award.

Although achieving great brands in the marketplace is the primary goal, it was very special to see the quality of P&G's **technical work** being recognized with the United States top award.

*"FOR SUSTAINED GLOBAL INNOVATIONS ACHIEVED THROUGH ACQUIRING A DEEP UNDERSTANDING OF WORLDWIDE CONSUMER NEEDS AND DESIRES AND TRANSLATING THESE THROUGH LEADING-EDGE TECHNOLOGY DEVELOPMENT INTO SUPERIOR HOUSEHOLD AND PERSONAL CARE PRODUCTS WHICH HAVE IMPROVED THE QUALITY OF LIFE AND HEALTH FOR ALL HUMANITY WHILE STRENGTHENING AMERICA'S GLOBAL COMPETITIVENESS"*

Six P&G products and the associated technologies were presented to the MEDAL OF TECHNOLOGY Award Committee, and were recognized for the award. They were:

**SHAMPOO-AND-CONDITIONER-IN-ONE....**PERT PLUS...When conventional shampoo and conditioners are combined, they interact and become non-functional. New conditioner agents were uniquely suspended apart from the shampoo surfactants in the product and in shampooing, and were released during dilution in rinsing.

**HIGH-PROTECTION FEMININE PADS.....**ALWAYS...The propriety "Dri-Weave"non-woven top sheet has unique tailored apertures which draw fluid into the core via capillary action, and prevent moisture from resurfacing. It keeps moisture away from the skin surface.

**TARTAR CONTROL TOOTHPASTE.....**CREST TARTAR CONTROL...The patented pyrophosphate formula disrupts tartar formation on the tooth surface. Any calcium that does form is not tightly bound, and can easily be brushed away.

**STRUCTURED TISSUE AND TOWELS.....BOUNTY PAPER TOWELS & CHARMIN BATH TISSUE.....**The process involves casting and selective ultraviolet curing of a liquid photopolymer film into a woven screen of fabric. The fabric forms a belt with a patterned surface that becomes the template for continuous paper sheet molding on a paper machine. The unique structures on the created film produce paper with superior absorbency, while maintaining strength and softness.

**DETERGENT WITH ACTIVATED BLEACH.....**TIDE WITH BLEACH DETERGENT.....The search for an effective fabric-safe bleach in a detergent formula had been a global industry goal for decades. The available routes were fraught with instability, fabric damage issues, and ineffectiveness. After years of study, P&G scientists decided that "activation" of peroxide in the wash to form a peracid bleach in solution was the preferred route. They conceptualized and invented a unique compound which had just the right effect to form the peracid, and to do it at the fabric surface without fabric or color damage. The compound was nonanoyloxybenzene sulfonate (NOBS). It was introduced as TIDE WITH BLEACH and other global brands in 1989, and became an instant consumer favorite.

**ULTRA-THIN DIAPERS........**PAMPERS...The biggest change in Pampers since its inception was made in 1986 with Ultra Pampers. They featured a patented Lock-Away core through the use of super absorbent, acrylic acid-based polymers (SAP), which acted as a replacement for some of the cellulose fluff and was fabricated into a compressed core. The ultra-thin core provided a 50% reduction in thickness, and offered superior dryness and absorbency.

I was very honored to receive the award from President Bill Clinton, and a full team of R&D managers and technologists attended the special award ceremony.

The news of the President's presentation of the award to Procter & Gamble was a surprise to many, but not to those who understood the P&G's focus on developing superior technology in their products, and for the strength of their Research & Development organization. It was a very fitting recognition of P&G's technical excellence.

### 2. THE COURAGE TO CONFRONT
### A COMPETITOR'S SERIOUS TECHNICAL ERROR

P&G always stood firm on its product claims, and was open to challenge competitor's claims when warranted. In 1994 there was a situation that occurred which was the premier test of P&G's technical reputation and quality of its work. It is simply identified as

# 'THE SOAP WAR"

between P&G and Unilever, a large European-based global consumer goods company.

Nabil Sakkab, a very experienced R&D Manager, was located at the P&G European Technical Center in Brussels, Belgium. He was in charge of the European laundry and cleaning business. His team had just received a product sample of PERSIL POWER, Unilever's laundry detergent, pro-claimed as the greatest advance in laundry detergents in history... behind a massive advertising campaign. When they opened the package they saw a distribution of small pink particles throughout the product.

Given our extensive knowledge of bleach activation technology, and awareness of Unilever's patents on the use of manganese compounds to activate bleaching performance, we

immediately concluded that Unilever was using this technology. It was an area that P&G had researched and abandoned because the technology damaged colored fabrics.

The team immediately launched a massive testing program of the Unilever product. They quickly saw the better cleaning performance of the product...but also documented the significant damage to colored fabrics after multiple washings. Not just fading, but major fiber weakening and holes in the garments. The manganese compound deposited on all fabrics, but was particularly attracted to the dyes in colored fabrics.

Ed Artzt, P&G's CEO, was on a London business visit a week later. He requested a full review of the Unilever product.

The meeting was held in London with a full contingent of P&G business and R&D management. Sakkab led a presentation of the findings. He and his team covered the technology, the positive bleaching performance as well as the major flaw in damage to colored fabrics. The team also showed that the damage could happen with P&G's ARIEL product, if it happened to be used after earlier washings with PERSIL or OMO POWER, the Unilever brands. Artzt asked Sakkab if he was **absolutely sure** of the reported findings! As all researchers know...."absolutely" is a benchmark they normally avoid...but Sakkab and his team were very knowledgeable of the chemistry. He answered 'YES'.

A call to Henkel, a second major competitor based in Germany, indicated that they had no information on the Unilever product or performance. At that point, P&G was alone in its assessment.

Mr. Artzt then proclaimed.....""We can't let this situation potentially damage our brand, Ariel.....much less the disruption to consumers with this fatal flaw. I want to talk to the head of Unilever now."

A call was made to Sir Michael Angus, Unilever chairman, who was on a sailing trip. Niall Fitzgerald, Detergent Category Coordinator - well known Unilever CEO candidate - was given the lead to handle the interface with P&G. He set up a meeting at the Unilever headquarters for the following day.

P&G had a contingent of seven people....Ed Artzt, Harald Einsmann, P&G European president, myself, Jim Johnson, General Counsel, Dr. Sakkab, Dr. Warren Haug -R&D Vice President, Global Laundry & Cleaning Products, and John Walker, research manager from our Newcastle, England Technical Center. Walker was leading the extensive laundry testing program on Unilever's products.

It was a "once-in-a-lifetime" scene with the seven P&G people slowly walking up the long front walk - think Gun Fight at the OK Corral - to the "Unilever House". Dozens of "eyes" were staring down on the P&G execs from seemingly every front window of the eight story building sizing up the enemy.

The meeting rather quickly evolved into two meetings. The R&D people from the two companies in one room…and the business executives in the other. P&G concerns and data were presented…and politely rejected by Unilever. They expressed great confidence in their technical position and people.

Artzt underlined that if Unilever couldn't resolve our concerns, P&G would go public with our findings. Additional technical exchanges were scheduled in Brussels at P&G's European Headquarters. We had hoped that with further work Unilever would acknowledge our findings. No such luck, our findings were irrelevant. The exchange didn't lead to any change of company positions. Unilever technical people knew their reputation was in serious danger; P&G technical people did as well.

Unilever slowed P&G's public efforts by suing and claiming P&G was defaming its products. Artzt brilliantly orchestrated the press response with "We welcome this lawsuit!…It will give us a chance to fully expose our data, and Persil Power's flaw."……Unilever shortly thereafter, withdrew the lawsuit.

However, a Holland advertising agency side-by-side comparison of ARIEL vs. POWER showing fabric holes was leaked. Media coverage literally exploded!!

From that day forward, the SOAP WAR was public…and almost no day went by without a story in some US or European newspaper or magazine. Most coverage positioned P&G as a "bully", trying to undermine a competitor's product. It was John Bull against the Uncle Sam upstart Cousins! It was high drama.

Things remained tense at P&G, as daily update meetings were held with Ed Artzt in Cincinnati, and obvious concerns about P&G's reputation had to be addressed with the Board of Directors.

P&G, however, established a terrific strategy for exposing the facts. We decided to enlist the various country consumer testing "institutes". These private testing institutes were very well regarded by consumers, were considered unbiased, performed high quality testing, and existed in almost every country. P&G shared their testing data with the institutes. The institutes bought into the need to do the testing themselves. In addition, P&G enlisted several private testing laboratories to conduct additional testing.

The public reports from these institutes occurring with consumers' real-life negative experiences reinforced P&G's position. In due course reports came from Holland, UK, Germany, Sweden, France and Spain all confirming the issue with the Unilever products. After about nine months Unilever publicly acknowledged the problem… and recalled their products.

It was a devastating, extremely expensive experience for Unilever. We were hoping the issue could have been resolved quietly - and took every step to allow that to occur. In the final denouement P&G's Ariel detergent gained consumer confidence, and grew strongly…while Unilever's European detergent suffered substantial losses. Guns back in the holsters!

Ed Artzt told me that shortly after when he was in London again, he got a call from the Unilever chairman, who wanted to come by to see Ed. He did….and his reason for the visit was to personally thank P&G for doing what P&G did to prevent an even much worse disaster.

The entire very unique story vividly illustrates the critical importance of the quality of R&D scientific work, and the confidence of P&G top leaders in their R&D organization.

## 3. RESOLVING MAJOR PRODUCT QUALITY ISSUES

P&G took the high ground publicly many times with its great scientific work. For example, the groundbreaking dental clinic work that brought CREST toothpaste the first endorsement established by the American Dental Association. Or the convincing safety studies that proved the allergenic safety of detergent enzymes under a global public microscope. Enzymes provided a major advance in laundry cleaning performance for global consumers.

 A particular situation I was personally involved in was the US national expansion of BOUNCE fabric softener in 1972. BOUNCE was a novel non-woven fabric sheet that was added to the clothes dryer for fabric softening and static control. Prior to launch it was thought to be a niche product that would take consumers substantial time to accept the required change from a rinse-added liquid softener product to a dryer-added sheet. However, when BOUNCE entered the market, sales unexpectedly rocketed forward. National expansion occurred quickly.

It was an exciting time, but soon the brand was being accused of blocking dryer vents, causing dryer fires. Then it was accused of removing the enameled paint surface of the dryers, causing rusting of the surface from the BOUNCE active, and subsequent damage to clothes. I was drafted from the Food Division to form a team to solve the problems.

These were very stressful times. Not only the difficult project challenges, but in this role, I was working for the first time (first of an eventual three times ) for Wahib Zaki (see Chapter 20). Wahib was newly in the US from Europe and new as the Division R&D director. I was adjusting to his highly demanding style, but also to his work schedule. He had what he called a European-style schedule, start time -10 am or so. He then worked until about 7pm. With all the issues we were facing, I was getting to work about 6 am. Zaki liked to engage with the team until he left. So, I was working regularly 12-13 hour days. Zaki was unmarried ....my wife and I had 3 young girls. Some days the extreme project challenges seemed the easy part!

The dryer vent blocking was solved quickly with a sheet design change. The Consumer Product Safety Commission was convinced with data we provided that the dryer fire accusations were not related to BOUNCE. However, dryer manufacturers remained EXTREMELY agitated with the dryer surface interaction.

The key manufacturers included all of the major brands: Whirlpool, GE, Maytag and Kenmore (Sears). They all used slightly different enamel paint systems, and all experienced some interaction with extensive use of BOUNCE.

The paint interaction and subsequent rusting was traced to a particular BOUNCE active, and with some brilliant chemistry work, a replacement compound was identified. The replacement turned out to be a very unlikely candidate - a unique food emulsifier - sorbitan monostearate. It was substituted because of its particular spreading and adherence to fabric properties, but also because of its absence of any interference with dryer surfaces.

It came from the brilliant understanding of surface chemistry by Paul Seiden, a Hungarian-born chemist in our corporate R&D group, with whom I previously worked with in the Foods area. Paul eventually achieved recognition into the Victor Mills Society, the top technologist award at P&G.

The reformulated product was thoroughly tested for performance with consumers, and then inserted into BOUNCE production.

However, dryer manufacturers indicated that they were going to recommend aggressively that consumers not use BOUNCE, and they would expose their concern about the product. They demanded proof that there was no interaction with the dryer surface for the **lifetime (avg. 15 years)** of the many dryer models.

How would we be able to generate the data for all the dryer models for their entire lifetime in any kind of reasonable timeframe? John Brock III, a young process engineer, and someone who later in his career became CEO of Coca-Coca Enterprises and the INBEV company, came forward with a plan. He suggested we buy all the largest laundromats in the Cincinnati area, replace all the commercial dryers with residential clothes dryers, and set up 24 hour-a-day operation with free laundry service. It would be realistic use of the dryers, and we could achieve the desired lifetime testing of the dryers.

Needless to say, a recommendation to the management to do this was about as radical as had ever been heard of. But it was put forward …. and it was agreed, and the plan was put into action. P&G was now in the laundromat business! The dryer manufacturers were engaged. They bought into the plan, and aided the plan by helping to install multiple surfaces inside the dryers, so that more than one surface would be tested in a given unit.

P&G employees were enlisted into using the facilities, in addition to local customers, and with the free service, home laundry cost in Cincinnati reached a new low.

The subsequent data reports were conclusive, supported the BOUNCE product, and BOUNCE went on to be a successful, respected brand. The laundromat experiment has to rank has one of the most out-of-the-box scientific studies in history, and certainly demonstrates the kind of creative, convincing scientific work that had to be done to save an important brand.

# 23.

# "MAKE A LITTLE, SELL A LITTLE, LEARN A LOT"

I'm a big believer in getting into the market as soon as possible.... normally a test market. That is, a limited market test of your product in full packaging, advertising, manufacturing, shipping and selling.

I've never seen a test market not yield important, usually unexpected learning. One just can't anticipate all the situations that happen in the marketplace in any kind of an orchestrated consumer test.

There's the chance to learn about weaknesses in the manufacturing, shipping, shelving and consumer experience - which happens in almost every test that's run.

Such tests have also resolved MAJOR doubts about the possible success of a product. Certainly, the market breakaway results from products like PERT PLUS, LIQUID TIDE, & FEBREZE all gave clear evidence of a major opportunity, and answered the very real skepticism that existed on these products prior to the test.

PERT PLUS- This was the first execution of P&G's breakthrough shampoo-and-conditioner in one. **ISSUE: The product would not survive the almost doubling of price versus existing shampoo brands, and would fail to deliver the high performance consumer expectations.** Instead, it rocketed to a 4X increase in the base share in 6 months. This pedigree experience resulted in quickly moving the technology to PANTENE, where it became the #1 hair care brand in the world.

LIQUID TIDE- The first ever change of product form for the Company's #1 brand. **ISSUE: The liquid product would not meet Tide's users expectations for superior performance, and seriously damage classic Tide's image.** The combined new product and package was a huge hit with all consumers, and received breakthrough ratings. The experience countered MAJOR reservations among P&G top management about any change to the Company's #1 brand. The product was expanded, built share, and within 15 years became Tide's #1 product form.

FEBREZE: A new brand launch in a difficult to differentiate, low profit, and crowded, low basis-for-entry category. **ISSUE: Is the Febreze proposition strong enough to justify a new, sustainable brand effort?** Febreze, the odor capturing liquid and spray exceeded every expectation… by a mile! It was an instant hit, with superior performance dealing with malodor. It exceeded anything consumers had previously experienced. They found unexpected new uses on fabrics, laundering and personal care. The broad acceptance and easy habit change projected a major market expansion opportunity. Febreze raced forward to become an iconic multi-billion dollar global brand.

There's always a risk with a test market that you will alert competition, and risk some market entry edge… or disclose a pivotal product or positioning. This concern came up with every product I was associated with. But I cannot recall any situation where, in the end, we weren't glad to have done the test. Unfortunately, I recall some where we wish we had - WOW - OLESTRA potato chips, BOUNCE fabric softener, among others. In those cases, we might have learned about the issues we came to face earlier than in national expansion, and worked to solve them in the test market.

Market Research companies, including P&G internal work, were actively working to develop test techniques that would fully mimic an in-market test, without actually needing to enter the market. While such tests have evolved over time, and while they were helpful, they never could acceptably capture the in-market learning.

Further, a prime reason to drive toward early market testing is

## "If you're going to fail, fail early."

Attached is a graph of the "typical" costs for launching a significant new brand from the start of the project… through initial market testing…and all the way through national expansion. It's easy to see that the costs early on in that process are quite modest. It's late in the cycle where the costs are very significant and a failure is a DISASTER. You just never, ever, want to fail late in the launch cycle. With Olestra, I know the pain.

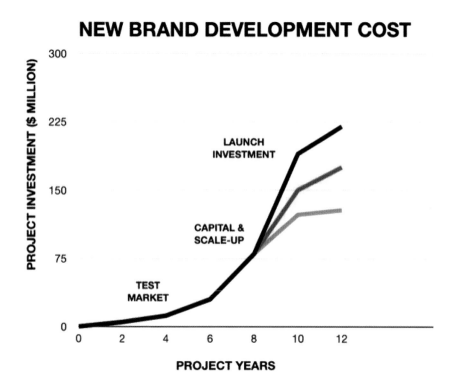

## NEW BRAND DEVELOPMENT COST

Early learning can lead to good project adjustments, and help pave the way to success. Of course it also can signal that the product is not going to be successful, and expansion can stop. This was the case in a new business with CITRUS HILL ORANGE JUICE. The innovative P&G freeze-concentration technology was not able to deliver target product quality in the test market. The high variability of starting orange juice flavor from the orchards had been underestimated, and the desired consistent superior fresh taste could not be achieved. The test was closed, and the business eventually sold.

The learning from such early market testing seemed to **always** speed up a successful development of a new product by the learning generated and the subsequent adjustments made.

I always firmly believed that generating big product ideas with good technology and getting early market exposure was a very sound financial investment, and it always was.

The upfront costs were modest compared to the long term major return that could come from delivering a major new brand. **Nothing wrong for trying and failing…learn a lot!…just fail early!**

And, as Henry Ford once said,

# "The only real mistake is the one from which we learn nothing"

# 24.

# "THE BIG DON'T EAT THE SMALL, THE FAST EAT THE SLOW"

In the innovation game, one learns pretty quickly the importance of speed. All the competitors are working hard to develop an important edge with the consumer, and working just as hard to do it quickly for maximum impact. Further, whenever P&G had an important product advance, getting that advance into the market, not only flawlessly, but quickly was paramount. Product ideas can be matched, proprietary advantages stolen, formulas back-engineered, consumer positioning and advertising ideas can be captured and boomeranged at you, quality glitches can stop you cold.

A clear example of the negative effects of speed issues can be seen in the 1960's in Europe. At that time, P&G was operating primarily on a country, rather than a united European basis. It was further complicated by duties that existed for shipping products across borders. But, at the time P&G didn't have any European plan in place as we introduced a new major product. It started with fabric softener, which was a novel new product introduced into Germany in 1963 as LENOR.(DOWNY in  US). It became the market leader with a dominant share. However, it was the second, third, or fourth brand introduced into the rest of

Europe. Henkel, Unilever, and Colgate were very fast followers, and adopted the LENOR plan before we could enter the other markets. LENOR did not match the German results in other countries in the early years of expansion.

The same pattern was repeated on MR. PROPRE (MR. CLEAN in US) household cleaner and Ariel detergent. The longer we took to expand across countries, the more difficult to replicate success. Eventually, in the late 1970's, a P&G united-European focus was put into place, with the planning and development of Euro-brands, not country brands, and Euro-expansion time eventually became a competitive advantage.

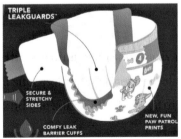

The critical importance of speed was highlighted with the development of LUVS disposable diaper. The basic design of a **shaped** diaper was developed by P&G researcher Ken Buell in 1962. This was a game-changing invention, and Ken later joined the inaugural VICTOR MILLS SOCIETY of top Company researchers.

At that time Pampers was just being expanded nationally in the US, plus P&G was getting ready for a major investment in a new paper-making process. The Paper Division was just earning its first profits, and the debate about how to use the shaped diaper was being debated with great pas-  sion across the Division. The cost of making the new diaper was 30% higher than PAMPERS Unfortunately, the technical position at the time was that there would likely **always** be a premium for a shaped diaper - which with further innovation longer term proved **not to be true**. However, that position, along with the financial challenges the Division was facing, led to the decision to market the shaped diaper as a premium priced LUVS brand.

LUVS was an instant success even with the premium pricing. P&G expected Kimberley Clark to follow with a flanker to their KIMBIES diaper, but that was wrong.

Instead KC pivoted and created a new shaped diaper brand HUGGIES, recapitalized their production capacity, priced the product equal to Pampers, and has given P&G a headache ever since. Consumers rapidly went for the shaped diaper. We had learned too lat**e to NEVER give a product reason for consumers to switch from a leading brand.**

We then compounded the bad strategic choice by failing to follow up quickly in Europe and Japan. It took another 25 years to expand ultra thin, highly absorbent, shaped diapers, into Europe.

Additionally, and particularly in the diaper business ,we learned one aspect of speed on a global basis the hard way. We came to realize the true punishing cost of **"being late because of being different."**

This is a perspective and learning that is not taught in any business school.. However, we learned the hard way about the true crippling effect of speed and costs incurred by the deviations from a strict standard for ingredients and process equipment, and also the manufacturing process layout. This was particularly true for a highly mechanical process like PAMPERS

For any number of strategic, financial, or organizational capability reasons, the original global expansion of P&G key brands took decades to accomplish.

- For example, PAMPERS diapers took about 40 years to have a presence in most parts of the world.

- Laundry detergents probably took more like 50 years.

- Hair and skin care products also on the order of 40 years

Because of how brands were originally gradually expanded to different geographies, decisions were made on the establishment of manufacturing and ingredient supplier selections based on the best costs available in that geographical location. If a key piece of fabrication equipment had a better payout for a given country, that 's what you did. Further, manufacturing equipment layouts were done to best fit the constraints of the building chosen for the manufacturing to be done.

As a result, with diapers being the most complex example, P&G had many different brands of key pieces of equipment - which originally accomplished the desired operation - but did it differently. There were many different suppliers of key fabric, film, tape, glue, etc. items in the diaper product, which originally accomplished the desired end result, but had different behaviors going through the equipment. There were a myriad of different layouts of equipment for producing a single diaper item- one single line….multiple side-by-side lines…lines which coupled together over multiple floors….lines which needed coupling between different buildings.

What's wrong about all that? In the end, the same, high performing diaper product was made.

Well, unfortunately, it was taking **multiple years** to get a new break-through design or product feature expanded globally. Why? Each plant was requiring its own team of technical experts to go through the qualification of materials, and process changes in order to qualify the new desired product. Different challenges were encountered in each location. They could be solved….but they took extremely valuable time. Time that caused lost business, and important competitive advantage in the marketplace. Of course, the substantial extra technical talent and time also created a penalizing cost. Further, P&G didn't have the substantial talent necessary to create an implementation team in every plant for the launch of a new technology. It had to be accomplished by moving people from location to location, again which took valuable time.

Speed was so important for the P&G diaper business, that once they understood the factors behind the punishing situation, P&G **recapitalized the entire diaper and feminine protection manufacturing facilities** in the late 1990's at a multi-100 million dollar price tag.. clearly a major strategic decision designed to achieve global speed.

As an aside, P&G had a brand that proba-bly represented the best, and maybe ulti-mate in expansion speed. That product was PRINGLE'S….

This was a brand that was manufactured in Jackson, Tennessee. It was the **identical** product in every part of the world. There was a broad lineup of flavors, but no other differences…. So, expansion was as fast as you could ship out of Tennessee!

But the overall learning is that key equipment , ingredient/material , and manufacturing infrastructure differences across the world can have important negative speed and cost effects on a global brand's innovation program.

As a result, expansion speed became a clear focus for improvement at P&G……A brand new hair care brand, PANTENE (Chapter **26**) was launched everywhere in the world within a year, and became the #1 hair care brand at the time. That was an amazing accomplishment versus the past, and will likely be a bench mark looking to continue to be bettered.

# 25.

# "ACHIEVING TARGET QUALITY IN THE MARKETPLACE IS CRITICAL"

## PRINGLES "NEW FANGLED" POTATO CHIPS

Since Bill spent significant time on this highly exciting innovation, he'll stroll through the story:

**BILL:**

In the late 1950's, the leading snack in the US, as it is today, is Potato Chips. Slices of potato, fried in oil, salted, & packed randomly in a bag. Potato Chips provided appealing taste for a lot of situations, but chips were excessively broken and not uniform for "dipping' situations. Oil flavor would oxidize quickly-often before purchase, chips were oily to touch, the bag was bulky in an attempt to create space to control breakage and it was difficult to seal after opening.

Corporate P&G researchers decided they would try to invent the perfect potato chip that eliminated all the problems. Researcher Fred Baur conceived a (hyperbolic paraboloid) saddle-shape for the chip, so they would stack compactly. He also conceived the tubular tennis-ball canister - a perfect choice for a stack of saddle-shaped chips. Fred was so proud of the can choice, that he requested that his children bury his cremated ashes in one of the cans when he passed - which they did.

(Hyperbolic paraboloid)

$$\frac{x^2}{a^2} - \frac{y^2}{b^2} = cz$$

 Alex Liepa, a brilliant Latvian engineer, led the design of the process. Potato flour (dried ground potatoes) was hydrated to form a dough. The dough was sheeted to a perfect thickness, and cut into "dovals", or the perfect shape to achieve the saddle-shape. The individual dovals were transferred and encased top and bottom into saddle-shaped individual carriers, fried in oil, salted, shingled to form stacks, and packed into the canister.

Importantly, Pringles could be distributed from warehouses, as opposed to direct store delivery with existing products. This provided a major cost and distribution flexibility advantage.

A pilot plant was constructed which accomplished the designed steps in a continuous process. A MARVELOUS achievement, helped importantly by similar work that had been done in the same Corporate group for high-speed production of Pampers diapers.

Consumer response to the chips was beyond "WOW"

A small production line was built at the nearby Cincinnati Ivorydale plant. The line was a larger version of the pilot plant. Once again, consumers loved the product!

A test market was quickly started. The market entry created an explosion of interest. The product was the toast of Evansville, Indiana for the media and for consumers. It clearly appeared that Pringles was a game-changer!

Given the continued highly favorable consumer experience, the Company authorized construction of a Pringle's plant in 1968.. It was to be located on a 50-acre plot in Jackson, Tennessee - 50 miles east of Memphis.

The Pringle's plant was going to be massive. An approx. 25+: 1 scale up of the test market line - a major Engineering challenge. The plant itself would have 6 production lines, and be large enough to handle US national expansion.

I had just completed leading a team to design a computer control system for the complex PRINGLE'S process, which we installed at the plant. I then returned to Cincinnati, and was now in charge of PRINGLE's Product Development, with national expansion ahead of us.

Competitors, seeing the popularity of PRINGLES, objected to the FDA that the product should not be allowed to be called "potato chips".

Extensive haggling over years led to PRINGLES labeled "made from dried potatoes", and later as "**potato crisps**".

The new plant lines started up well mechanically. The mammoth operation was somewhat of a spectacle to watch as speeds were gradually ramped up. Giant 50-ton roller mills pressed the potato dough into the desired thickness, and saddle-shaped PRINGLES moved through the frying oil at 100 Ft./min. A one-of-a-kind, space-age type, full-scale manufacturing process was producing products soon to be in the market broadly for the first time.

But, an issue emerged. The PRINGLES tasted "dry and grainy", and lacked the richness and potato flavor of the test market target. However, they were judged to be at the low end of an "acceptable" range, and it was felt that further lining out of the new process would correct the issue. Further, every possible PRINGLE'S can was absolutely needed to meet the amazing market demand. The market expansion was going so well that a second plant in Greenville, North Carolina was authorized.

But behind the increasing production figures was a warning sign. Sales in the first expansion areas were declining. The euphoria of the first production chips gave way to the realization of a product deficient in flavor and texture. Consumers were reducing their purchases and, in many cases, stopping them. Pringles had a problem.

Every possible modification of the plant was going to be tried in order to bring the PRINGLE's product to target quality. This was urgent, because we could see a slow decline in PRINGLE's volume.

At that time, I received a promotion into the Coffee Division, and left PRINGLE's I lost track of the PRINGLE'S situation, but returned to the Food Division seven years later as Food R&D Director. I had been aware of PRINGLE's issues, but I was quite shocked that PRINGLE's had collapsed. Volume was down 60% from its 10MM case peak. Greenville, NC was built, started up, then mothballed. The brand had become a niche curiosity.

The product quality situation remained an issue as before. The taste deficiency had not been fixed. **The Jackson plant design could not make target product**.

In my first Product Review with US President, John Smale, I got my PRINGLES marching order:

> *"There are several of your Finance, Sales and Advertising colleagues who think we should stop Pringles production and write it off. Now most of our costs are sunk-costs and our operating losses are small. So, there isn't too much of a financial burden to continue. This is a large and attractive food category. Our basic premise for a superior stackable potato chip still seems strong. Bill, I want you and your development group to do everything you can to make PRINGLE'S a successful brand."*

We had been working hard to do just that. One of our culinary all-stars, Yen Hsieh was working on adding different flavorings to the surface of the chips, right out of the fryer. They tasted great. The flavor solids added onto the somewhat bland Pringle created a great flavor experience. In the Pringle's canister, we avoided the flavor deterioration of the added flavors seen with conventional bag products. With a minimum of consumer testing, we launched Cheddar Cheese PRINGLE's nationally as a flanker product.

**It was an immediate success…consumers loved it!**

Cheddar Cheese PRINGLE's added 1.0 MM cases ( with virtually no canabalization). The Team quickly added a Sour Cream and Onion flavor with good success, adding an additional 600 M cases. The flavored Pringles had started what turned out to be a palette of flavors, and a great product platform  for PRINGLE'S There are now 23 in the marketplace and an astounding 166 flavors over the past 40 years.

I had moved off PRINGLE'S, but in the late 1980's John Smale gave a similar mandate to Bob Gill, incoming VP on the brand,"make Pringles profitable by filling the plant". At the same time, Chuck Hong , an experienced researcher, led a team who discovered the addition of different grain flours and malto-dextrin (a natural starch) enhanced the over-  all flavor in an unpredicted manner. Because PRINGLE'S was now "potato crisps", these food ingredients could now be added.

Gill formed a multi-disciplined team from R&D, Engineering, Manufacturing & Marketing located at the plant, and took a 'no-holds barred" approach to attacking every element of the brand.

The improved product and flavor versions, along with improved marketing, drove the business for-  ward. In seven years, not only was plant capacity filled, but a 60% increase in line speed resulted in a 40% decrease in manufacturing costs.

Demand exceeded supply for the first time. The Jackson plant was expanded, and a second plant was constructed in Mechelen, Belgium and then Malaysia to satisfy global demand. Specific flavors were added to satisfy local interests. **By the end of the 1990"s, PRINGLE's had reached the "BILLION-DOLLAR BRAND" level at P&G - a terrific achievement! The brand had grown over 15X from its low point.**

Later, as part of P&G's desire to exit the Food & Beverage business categories, PRINGLES was sold to the Kellogg Company in 2011 for a reported $2.7 billion

One can only guess what if the original taste quality of Pringles had been delivered at the start? What would have happened if the negative image of Pringles taste had never been established? Could Pringle's have been the leading global potato chip brand? .......We'll never know.

There are several conclusions that can be drawn from the PRINGLES experience:

1.  Process scale-up for delivering target product to the market is critical.. It's a particular challenge for complex mechanical and chemical systems. While there are a variety of possible explanations why it occurred on PRINGLE's…it's clear a critical factor was missed.

2.  John Smale's pointed effort to engage and provide clear "keep going" messages was motivating and critical. It again points out the critical effect that happens when the "decision maker" engages deeply with the innovation work.

3.  The "turn around' of PRINGLES with the base product improvement and flavor extensions is just another great

example of motivated researchers not giving up, and coming up with the needed answer to solve the challenge. It's been said,

# "It only takes one idea to solve an impossible problem"

# 26.

# MAJOR INNOVATIONS COMING TO LIFE BEHIND BIG PRODUCT EDGE

The stories behind the creation of major product innovations need to be memorialized, as they represent precious stories of committed researchers overcoming the inevitable daunting barriers of "changing the game". The resulting products create delight for consumers, and long lasting great rewards for shareholders. But, every great innovation story also provides learning to be reapplied, and encouragement for other development teams as they pursue their own innovation quest.

In the 1980's, behind the drive for 'BIG PRODUCT EDGE" , and the major personal product innova-tion focus from CEO John Smale, ALWAYS "Dri-Weave" Feminine Pads, PAMPERS Ultra-Thin Diapers, and TARTAR CONTROL CREST, which were all part of the 1995 US MEDAL OF TECHNOLOGY AWARD (CHAPTER 22) were successfully launched.

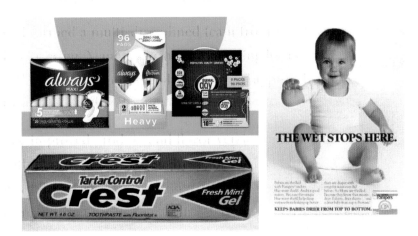

Further, I'll now cover the stories on the creation of the following additional major product innovations which launched in the 1980's:

- WORLD LAUNDRY LIQUID - TIDE & ARIEL LIQIID

- WORLD LAUNDRY GRANULE - TIDE with BLEACH

- PANTENE HAIR CARE

- STRUCTURED TISSUE AND TOWELS (CHARMIN & BOUNTY)

- ACTONEL FOR OSTEOPOROSIS

- FLEXIBLE ABSORBENT MATERIAL - (ALWAYS "INFINITY")

# "WORLD LAUNDRY LIQUID"
## ~TIDE & ARIEL LIQUID~

Liquid laundry detergents started appearing in the market in the US in the mid-1950's. Unilever introduced WISK but it did not have the cleaning performance of laundry powders, and was used primarily as a pre-treater of stains. P&G top management made it clear to the organization, that we **never** were going to introduce a liquid version of TIDE that did not have the cleaning performance of TIDE powder.

P&G did introduce ERA liquid in 1972, which was primarily used as a stain pre-treater and SOLO liquid in 1975 as a detergent plus softener. Both struggled for market share.

The existing technology for liquids at the time was primarily a dilution of the powder formula. As a result there were powder solubility issues which prevented desired concentration, but primarily, with the liquid formula and high pH, critical detergent stain-removal enzymes were neither stable, nor effective.

An effective laundry liquid required a new basic structure. Beginning in the early 1960's, P&G put its energy behind a new "builder" - the prime ingredient for neutralizing water hardness in a detergent, and which for decades was provided by phosphates. The compound was NTA (sodium nitrilotriacetate). Twenty years of extensive development and safety testing was done, only to have the US EPA in 1980 **deny** approval for extremely conservative concern over NTA byproducts and environmental decomposition. Over $200 million dollars had been spent on the product development, safety and environmental studies, supplier commitments, and capital expenses for plant modifications behind the NTA technology. **A truly major setback!**

 Meanwhile, WISK had built US share to 10%, behind a "Ring Around The Collar" positioning, and along with a SURF powder new brand launch behind a highly preferred fragrance feature, Unilever was taking share from TIDE. TIDE'S share had sunk to its lowest level in 25 years.

It was a very concerning time. The technology strategy behind NTA had blown up, and there was no path in hand to counter TIDE's market issues.

John Smale, who had just become CEO in 1981, was unhappy with the US laundry market situation, and also the overall R&D new product productivity. He was impressed with Wahib Zaki (Chapter **20**), the European R&D head, from both the US Bounce experience, and also seeing what had been done in Europe with Euro-products, and a turn-around in European profitability . He asked Zaki to become CTO reporting to him. He stressed the great urgency on US TIDE.

Having gone through uniting effort behind Euro-products, Zaki decided to form Global teams on World Laundry Granule, World Laundry Liquid, and World Fabric Softener. That is, all the R&D global resources would be focused on a common objective - breakthrough laundry products for every market in the world.

I had taken Zaki's job as European R&D head. He asked me to lead the World Laundry projects. After about a year he asked me to come to the US in a Vice-President role, retaining responsibility for the World Projects. Dr. Claude Mancel in Europe, Dr. Warren Haug in the US, and Doug Moeser, for Latin America, Japan, and Canada, headed up the different R&D groups who were fully engaged in the development effort.

All the World Projects were eventually successful technically and in the global markets. But World Liquid had a truly major effect, and I'll recount the story.

Liquid laundry product technology progress had actually started in Europe with Mancel's team some 6 years earlier. Their team had major input from Jean Wevers, a European researcher who later became recognized as member of P&G's Victor Mills Honorary Technologist Society. He developed the concept for laundry liquids to focus on maximizing the use and effectiveness of detergent enzymes. Detergent enzyme technology was continuing to advance, with further improvements from suppliers expected. So it was a strategy that also had longer term strength. However maximum enzyme performance required the product to be neutral pH - a very

difficult challenge for effective detergent cleaning. To achieve the neutral pH, a combination of organic fatty acids were used as the "builder". This approach was very novel, and this European strategy formed the basic structure that was used to advance WORLD LIQUID.

While the global work got underway, Europe separately launched test markets of the base technology in a new liquid detergent brand, VIZIR, which had been an active project before the WORLD LIQUID project was launched. The test market experience was very helpful, because it not only gave early experience with the base formula, but also identified a "sump loss" problem. In European washing machines, some liquid went straight to the machine drain "sump" before starting, and was lost. To correct this problem the Euro Team designed a small spherical device for adding the liquid to the machine. Called 'VIZIRETTE', it prevented loss, actually pretreated the clothes at the start of the cycle, and was a real hit with consumers.

Over the previous several years, a great deal of development work had been done with global detergent enzyme suppliers. A line-up of new protease, amylase, lipase, and cellulase enzymes now offered the opportunity for a major advance in stain and soil removal. A new superior combination of surfactants and soil suspension polymers added to the formula's strength. Finally, the neutral pH formula added fabric color and fiber safety advantages.

P&G's packaging group created a brilliant design for a "self-draining" bottle cap thereby eliminating the messy drips after measuring. This was a real negative and nuisance with current products - and the bottle design received patent approval.

Further, for the first time, the label was encased inside the plastic bottle wall, a great improvement in appearance and a significant simplification in manufacturing.

So, we had a great detergent liquid formula, and a superior package… NOW was it good enough to be called TIDE? Our consumer testing of the product and package showed outstanding consumer acceptance. Although the liquid formula was entirely different than the powder formula, and the mechanism of cleaning was different, the overall cleaning performance of the liquid was at least as good as the powder.

The European and other country consumer testing of Ariel Liquid was at least as strong. The global technical group now was clearly of the mind that this World Liquid product should be marketed as TIDE LIQUID & ARIEL LIQUID.

I went to Wahib, and urged him to try to sell this course with John Smale. He did, and returned with a quite firm **"THIS JUST ISN'T GOING TO HAPPEN".** After all this great technical work, although not completely unexpected, Smale's response was totally deflating!

Fortunately, we had a scheduled annual product review coming up with John Smale. In the meeting we thoroughly covered all the positive consumer and laundry testing we had on World Liquid. I concluded: "John,,,this product clearly deserves to be called LIQUID TIDE." He frowned disapprovingly…and then we dealt with his great concern. 'Isn't this going to make it easier for competition to enter the market without spray-drying?" The team's response, "Agglomeration is advancing so well that spray-drying won't be required any more for powders ( which happened)…In addition, several of the new high-performing ingredients in World Liquid won't work in a spray-drying tower. The World Liquid formula and package is so strong it will be substantially more difficult for anyone to close the gap - EVER.. (which has been true for 35 years to date)"

There also was concern for the higher Liquid Tide cost vs. TIDE powder. We simply said we would eliminate the difference in six months (and did).

The entire global WORLD LIQUID team was united on wanting to launch the product as LIQUID TIDE & ARIEL. A recommendation to do parallel test markets of LIQUID TIDE and a new brand was made. John thankfully agreed, and global cheers resounded.

The LIQUID TIDE test market quickly was a **roaring success** … consumers loved the product.!

It didn't take long for liquids to become the preferred form for laundry detergents worldwide, and TIDE LIQUID raced by 60% share of the brand in the US market. As the market today moves on to unit-dose liquid-containing pods, the fundamental technology behind LIQUID TIDE and LIQUID ARIEL continues to drive consumer performance and acceptance.

US TIDE'S share rebounded, and kept growing. Under continuing product innovation pressure, Unilever eventually **sold** their US detergent business in 2008. Certainly a resounding win for TIDE. And equally exciting, Liquid Ariel was a market success in every country it entered.

## GLOBAL LAUNDRY LIQUID truly became the liquid laundry detergent for the world..

# "WORLD LAUNDRY GRANULE"

Alongside the World Laundry Liquid project, a separate team was focused on a significant advance in global laundry granule products, in a project dubbed WORLD GRANULE.

The project team had access to the same advancements in enzyme, surfactant, and suspending polymers that were being used in the WORLD LIQUID project, but the primary additional focus was on the development of a breakthrough oxygen bleach.

Efforts at a superior laundry bleach had been a goal of laundry detergent competitors for decades. Sodium perborate or sodium percarbonate were oxygen bleach sources that were broadly used in European detergents, where wash temperatures are often above 60 deg.C. Under these conditions, hydrogen peroxide bleach was released in the wash.

This approach was not effective in the rest of the world where wash temperatures are much lower.

Various metal catalysts can form complexes with oxygen bleach sources that are effective, but have disastrous side effects (see "soap war"-Chapter **20**). A preferred way was to use an activator that interacted with the oxygen bleach source and safely produced a peracid in the wash, which was more effective than the oxygen bleach. Such an activator had been identified earlier in Europe called TAED (tetraacetyl ethylene diamine), and was being used. However, it still was not effective outside of Europe.

 So, the focus was to develop a superior activator for the world….and a major chemistry invention was needed! This work was led at the Newcastle, England technical center. Dr. Fred Hardy, a world class chemist, who had spent most of his career studying oxygen bleach reactions, finally came up with a breakthrough answer.

The discovery was the compound called NOBS (nonanoyloxy benzene sulfonate). It produced a more effective peracid in the wash solution, and worked at all wash temperatures. It formed a key proprietary technology underpinning WORLD GRANULE. Fred's career work led to his inaugural entry into the Victor Mills Honorary Technologist Society.

It was decided to launch the formula as a separate item in the US., as TIDE WITH BLEACH. A test market was launched, created immediate consumer interest and approval, and was expanded.

WORLD GRANULE formulations were also developed for the complete line-up of global detergents. They produced global consumer acceptance advances, and were launched with broad market gains. The very first time a major advance had been made on the total global heart of P&G's business....in a single innovation!

# NEW BRAND
## ~PANTENE~

In 1980, there was seldom a clearer consumer need expressed in any category than what existed in hair care products. Consumers were shampooing more frequently, women's hair styles were longer, and separate hair conditioners were not used or of poor performance. Hair was overly dry, frizzy, unruly, and damaged. Consumers were very clear on the need for a shampoo with hair conditioning benefits.

However, combining shampoo and conditioner in one was very difficult. With given technologies, the two would interact chemically, and the resulting product wouldn't clean or condition the hair.

P&G, at the time, had a successful brand in Head & Shoulders, an anti-dandruff shampoo. However, it had been unsuccessful in developing a leading brand behind SHASTA, DRENE, PRELL, IVORY, & PERT past national US launches, and a couple other test market attempts. A very poor record that provided unspoken, but deep angst and energy among the hair care researchers to rectify the past.

Under CEO John Smale's strong urging, it was decided to make a big push to develop a shampoo & conditioner in one. Ceil Kuzma was R&D director in US Beauty Care, and partnered. with Franco Spadini, a R&D director in corporate R&D. Ceil was a researcher who understood consumer's hair care needs, and in a past product review had first convinced John Smale that P&G **had** to get into hair conditioners, even though there was no internal commercial interest at the time. Franco was a highly talented Italian-educated researcher who came to the US from P&G's European Technical Center. The project was labeled BC-18.

A number of unique technical approaches were identified and tried - and failed. Finally, it was Ray Bolich, who later became a Victor Mills Honorary Society member, who identified a unique approach. He chose a then new hair conditioning material, polydimethylsiloxane, - a silicone derivative, as the conditioning agent. The idea would be to form very small particles of this material in the product, suspend the particles in the shampoo, have them be stable in the bottle and in the high concentration of shampooing…..and then deposit on the hair during the hair rinsing dilution.. Sounds impossible? Well, the total team went to work, and brought it to reality . The proprietary possibilities were extensive.

Consumer testing moved forward. It proved difficult because of the unexpected benefits consumers saw, and didn't quite know how to react to in a blind product situation. But, the unique product characteristics were definitely delivered.

As the decision was getting close on proceeding to test market, the Beauty Care Division President had decided to **not proceed** with the project. He had high concern about needing to premium price over a normal shampoo, uncertainty about how to brand and position the product, and had significant profit pressure within the Division, which had nothing to do with BC-18.

The BC-18 team was crushed. I was the R&D Vice-President involved with hair care among my responsibilities. I was VERY mad as well. I decided I would go right to CEO John Smale and make the case. I knew it was a waste of time trying to change the Division President's decision. I bypassed him and also my boss, and made an appointment with Smale. In hindsight, it was probably pretty reckless. I had previously only interacted with Smale in the Bounce experience, and a meeting in Europe…but he remembered me. Our meeting was brief. I merely highlighted that "it was a breakthrough product,…a significant invention…. and we really needed his help to get a chance." He said he would consider it.

What I didn't know is that he immediately walked down to Bill Connell's office, who was in charge of hair care, and told him to get it into the market. The plan was to take PERT, a brand that was positioned on a fresh fragrance, and was struggling, and launch it as PERT PLUS. A friend, Hank Feeley, who was  vice-chairman of the Leo Burnett Ad Agency and his people developed the idea of a "Wash & Go" shampoo. The test market was in Seattle,  Washington. I urged Bill to "sample the heck out of it"…which he did. The former struggling brand rocketed forward. Consumers loved the product! We knew we had a winner! Plans proceeded to expand PERT PLUS globally.

In the meantime, from the acquisition of the Richardson-Vicks Company, P&G now also had the ViDAL SASSOON hair care line, and a department store brand based on vitamin infusion, called PANTENE.

Ed Artzt was in charge of International, and in globalizing the Company. He was very aware of the PERT PLUS results, but felt the "Wash & Go" positioning was not a winning global brand strategy. He had studied the PANTENE heritage and story, and saw the brand had a wonderful premium image broadly, even though its very high department store price and limited retail distribution prevented major volume. The vitamin penetration to hair was a strong consumer concept, and nicely fit the BC-18 product benefits.

Artzt challenged several country teams to try to develop a winning global brand concept behind the PANTENE name and BC-18 product. James Wei, the country manager in Taiwan, creatively took a Japanese advertising campaign behind **"Hair so healthy it shines"**, terrific Japanese advertising visuals showing beautiful shiny hair, Japanese PANTENE package design, along with a full line of shampoo and conditioner products,,,,,started a test market.....and found a roaring success!. It became the model for the PANTENE global brand.

Artzt challenged the global hair care organization to achieve global expansion within a year. This was a HUGE stretch beyond anything that had previously been achieved in the Company. With all the challenges on every aspect of the product's expansion in every global market, it seemed only wishful thinking. But the expansion in 32 global markets was completed not in a year, but in 10 months, and **PANTENE was the number one hair care brand in the world.!**

A great global product innovation story with the US, Taiwan and Japan providing the innovative pieces, and all the global markets providing the execution speed for success.

# -MAJOR INNOVATION -
# STRUCTURED TISSUE &
# PAPER TOWELS

P&G had gotten into the paper business by acquiring the small Charmin paper company in 1957. A breakthrough paper-making technology evolved involving blow-through drying. It was termed internally as the CPF project. Instead of removing water from the wood fiber/cellulose slurry by squeezing out the moisture under tons of hydraulic pressure, the wood fiber mix was subjected to high pressure air drying. The resulting tissue was softer and more absorbent, and consumer testing of facial and bath tissue products showed great consumer excitement.

But, in the paper business, scale-up of ideas to production scale is every bit as hard, and certainly vastly more expensive, than the fundamental invention itself. In retired P&G CEO Ed Artzt's oral history book ( "The Globalization Years, Gatekeeper press, 2023), he detailed the challenges of commercializing CPF, when he was Paper Advertising Manager in 1965. I'll briefly summarize: what he outlined: CPF was the installation of Blow-Through Drying, which permitted effective water removal from the woodpulp slurry prior to the paper machine. It avoided much of the high pressure compacting of the fibers, yielding softer, more absorbent and consumer-preferred paper, and greater fiber use efficiency.. Howard Morgens, the P&G CEO, knew that CPF was essential for the P&G Paper business to survive - to create the competitive product advantages and the necessary cost structure. But over $300 million had been spent on the scale-up, and so far it had been unsuccessful. Morgens personally intervened, and put his eventual replacement, Ed Harness in charge of the project, with a blank check on people selection and capital spending. The top Engineering and Manufacturing technical talent, including Harry Tecklenberg, who later became Chief R&D Officer, were assembled, working with Ed Harness. The technical innovation work and the top decision makers were now fully linked. A very tense followed, but with brilliant technical work, it finally yielded success.

The CPF invention permitted major market advances for the original CHARMIN tissue product, along with PUFFS facial tissue and BOUNTY towels. P&G was now clearly viable in the Paper business.

 In 1985, Paul Trokhan, a seasoned paper researcher and future inaugural Victor Mills Society medalist, had a string of patents issue behind a major advance in the same critical paper properties that CPF originally accomplished. The new process involved casting and selective ultraviolet curing of a liquid photopolymer film onto a woven screen of fabric. The screen of fabric forms a belt used for paper sheet molding when mounted on a paper machine. A typical belt was several yards wide, more than 100 feet long, and at the time, cost many tens of thousands of dollars each to produce.

The process was a conceptual breakthrough, as it provided essentially unlimited patterned surfaces for major modifications of paper properties. However, it had to withstand highly challenging chemical and physical forces on the paper machine. Paper is created at a speed of 2,000 ft/min, temperatures exceeding 500 deg. Fahrenheit, and pressures up to 10,000 lbs./sq.in. For this story, I'll refer to the new process as NP-1

When I became CTO in 1987, the NP-1 technology had been under development in a plant environment for a couple years to demonstrate viability. Tens of millions of dollars had been invested to date, and it had not yet proven feasible. The NP-1 belt would break prematurely. For viability it needed to run for many days, but in the last run, it broke in a handful of minutes.

John Smale had backed the project fully because it was a true breakthrough - the kind of project which he consistently challenged the organization to develop. However, he was losing patience, seeing milestones consistently missed, investment costs mounting, and significant concerns mounting from the CFO, and board members.

My previous experience in paper was only with Pampers in Europe, and John was aware of that. But he asked me to get involved with the project, and give him my point of view. After studying the history and numerous files and reports, it culminated with a meeting with all the key product development and manufacturing people. I was quickly struck by the fact that this was one of the most talented, experienced, and passionate technical groups I had ever experienced - and that covers some amazingly talented teams.

 Bob Haxby was the NP-1 project leader, and Paul Trokhan the key technologist. The team's understanding of the basic engineering issues and technology of a new structure that had never been done before was impressive. Although they had a lot of setbacks, many of the challenging technical issues couldn't be simulated in the pilot plant. It had to be done in the plant. With each setback, the team came away with new learning that built confidence they could achieve the goal of a structured paper/towel product. . Based primarily on the quality of the people, but also realizing there would be further setbacks, I indicated to John that we should stick with it.

The benefits were **game-changing.** Fundamentally, the technology gave unbounded ability to alter strength, softness and absorbency of the paper, in harmony for a particular desired end result. **It could produce paper characteristics and benefits never before achievable**

The project next came to a head when Haxby requested another $10 million to continue. In a subsequent high level meeting with all of the key players, John Smale expressed that he didn't feel he could support the request given the problems and costs that had been experienced. Haxby, looking straight at Smale, said, "Trust us John…we will deliver this…

## just trust us…don't give up".

With a slight pause, which seemed forever to the NP-1 team, John indicated."OK…We'll write the check"

With additional work, the NP-1 process was proven feasible. Haxby's team came through. The benefits to the Company were phenomenal! In the10 subsequent years, BOUNTY towels grew from 10 million cases to 42 million, and share doubled to over 40%. CHARMIN toilet tissue share grew by 40%. Additionally, and very important, both brands experienced major growth in profit margins.

John Smale's personal close involvement as CEO was pivotal. His willingness to continue to back this team and this breakthrough technology created the basis for major success in the paper business for decades into the future. It's an amazing example of the importance (and difficulty) of major product innovation, and the value of close linkage of the technical work with the decision maker.

# NEW BRAND
# ACTONEL Rx

The interest in pursuing a drug to counter osteoporosis, or the weakening of bones from undesirable bone resorption, stemmed from the research work of Marion D.(Dave) Francis. Dave was a Canadian, who received his PhD in chemistry from the University of Iowa. He became broadly recognized for his work at P&G, and was also an inaugural  member of the Victor Mills Honorary Technologist Society. He was awarded the Perkin Medal in 1996, an annual award presented by the Society of Chemical Industry to a single individual. It is the highest award given for excellence in Chemistry in the United States, and Dave was the first and only member of P&G to ever receive the award.

Dave spent his career making critical discoveries on the effects of calcium in P&G products. Following his academic training, he worked on calcium control in detergent applications in his first assignment. He moved into the dental area, and was key in proving the positive effects of fluoride on calcium loss from dental enamel, as well as the key mechanism for tooth tartar removal behind TARTAR CONTROL CREST

In the course of his work, he was the key researcher to understand and project the positive effects of bisphosphonates in preventing both bone loss in osteoporosis, or poor bone formation in Paget's disease. This led to the patenting of a specific bisphosphonate drug candidate, risedronate.

Based on this finding, and P&G's deep understanding of calcium control from their detergent and dental history, Harry Tecklenberg, then Chief R&D Officer, urged the Company to move into the Pharmaceutical space to pursue risedronate's development. The Company purchased Norwich-Eaton Pharmaceuticals in 1982, and expanded their internal pharmaceutical R&D capabilities as well.

The decision was made to initially pursue the development of risedronate for treatment of post-menopausal osteoporosis. This represented a huge consumer need and potential major market opportunity in modulating the debilitating effects of this disease. Extensive safety and efficacy studies were conducted over the next 13 years, prior to formally proceeding with extensive studies seeking FDA approval.

The existing FDA protocol for osteoporosis drug effectiveness was to generate proof for reducing spinal bone fractures in women of high fracture risk. This clearly was a very challenging goal, with major time and cost demands. Further, the variability in bone strength among participants meant very large test panels to eliminate bias, requiring accessing extensive clinical test capabilities.

Meanwhile, Merck Pharmaceuticals acquired a bisphosphonate compound from Instituto Gentili, an Italian pharmaceutical company. - called alendronate. With their massive capabilities, Merck was able to move through FDA clearance for postmenopausal osteoporosis in record time, and went on the market in 1995 with the product FOSAMAX. P&G was 2-3 years behind, if everything went well. Merck, which had now established themselves with the FDA as the premier authority for osteoporosis, was feeding perspective to the FDA fo requiring additional test criteria in osteoporosis clinicals. This added to the time, cost, and risk to the P&G clinical studies. Clearly, Merck did not want to see P&G enter the market.

Harry Tecklenberg was in charge of the Pharmaceutical business, and Gil Cloyd was the R&D Vice President. In total, eight demanding clinical studies were run, with over 10,000 women involved. The total investment and extreme risk was major, and something P&G had not experienced before. This wasn't one horse among many horses to be run. This was THE single P&G horse. As the time closed before the results would be available, the tension got major - both among the executive team and in the board room. But, in spite of the fact that P&G was entirely new to the game…..the clinical studies were successful. Actonel was approved for postmenopausal osteoporosis in 1998, and was launched as ACTONEL.

Osteoporosis was a common concern for all postmenopausal women when bone protective estrogen levels lowered, and bone loss became a clear possibility. Bisphosphonate treatment became widely accepted, and prescriptions soared. ACTONEL received approval for male osteoporosis in 2006, and for Paget's disease in 2008. ACTONEL had become a very successful prescription medication.

As P&G went through consolidating their business categories, and facing major investment and commitment to remain in the pharmaceutical business, they decided to exit the business. They sold the entire pharmaceutical business, with ACTONEL the main asset, to a leading women's health pharmaceutical company, Warner-Chilcott, in 2009 for $3.1 billion. Warner-Chilcott was later bought by Allergan Pharmaceuticals.

Several years later, Merck came under attack with accusations of osteonecrosis of the jaw, femur fractures and esophageal cancer associated with long term use of FOSAMAX. Thousands of lawsuits were filed against Merck, and bisphosphonate was selected as a lawsuit favorite by law firms seeking a large group of common issues. Any incidences were estimated as 1 in 100,000, but the bisphosphonate category became negatively tainted , and prescriptions of ACTONEL were severely affected.

Overall, ACTONEL serves as a major example of leveraging and betting on a company's technical competency. It involved the very highest risk, which pharmaceutical drug development involves, and yielded success. But, it also included the other aspect of reality in pharmaceutical drug development - every drug is a balance of positive and negative effects. Understanding the negative effects is usually more difficult than understanding the positive effects. But that understanding is equally critical, poses a definite liability, and always needs to be an open alert to the physician and the patient.

# -BREAKTHROUGH TECHNOLOGY-
# FLEXIBLE ABSORBENT MATERIAL
# (FLEXFOAM)

For many decades since the discovery of plastic-producing polymers, researchers have been working to create different types of foams or sponges. The idea was to create products with a range of densities, flexibility, absorbency, thickness, and other properties, for a myriad of applications. By varying the chemical and the process, products were created for construction, sound absorption, shielding electromagnetic interference, and all types of padding and protective applications. Countless forms were created - blocks, pellets, custom shapes, sheets, and many others.

In the late-1980's, the idea was generated by researcher Tom Desmarais to create a flexible foam material (FlexFoam) for superior performance in diaper and other absorbent situations. It would not only be highly absorbent, but superior in liquid retention, and highly flexible.

The idea would be that its internal structure would not deform randomly upon liquid entry as with then current pulp and AGM absorbent material. It would be preformed for consistent and designed performance, and providing an entirely new level of absorbency and comfort.

The technology was complex. It required taking a tailored emulsion of polyacrylate through process and drying steps to achieve the desired cell structure, while maintaining flexibility and resilience to not loosing shape. It was beyond the know-how of anyone in the industry. Cost projections suggested it would never be practical.

Some progress was made, but the project ground to a halt as the challenges didn't yield.

 In the late 1990's, I commissioned a new corporate organization focused on "New Platform Technologies" - those technologies potentially applicable to a broad range of P&G's businesses. It was headed up by Jeff Hamner, an experienced researcher with a very strong "can do" attitude. Among several major new ideas, I urged we tackle FlexFoam..

Very significant progress was made, but focus was shifted from diapers to feminine care. FlexFoam was not suited for a large surge of liquid, as occurs with diapers.

Alvaro Restrepo was the head of the feminine hygiene R&D group. He concluded that the existing category technology was not going to be robust enough to maintain ALWAYS superiority in the future. He convinced the president of the category, Melanie Healey, to apply the bulk of category technical resources against the FlexFoam technology.

ALWAYS prototypes were constructed with a unique benefit of incredible fluid absorbency as well as distribution and retention through a layered FlexFoam approach. The benefits were performance and comfort. (ZERO LEAKS & ZERO FEEL)… Consumer response to early prototypes was WOW+! -with an experience of performance and comfort they have never felt before.

After additional progress, and although major scale-up issues loomed, Restrepo convinced Healey to move forward with a $100MM plant investment. Once started, major scale-up issues ranked with the most ever seen in other Company paper product projects, and the project was almost stopped.

But progress was made. Major issues on cost and FlexFoam production feasibility remained, but the consumer reaction had been SO positive, that Steve Bishop, the new category president, and Restrepo, decided they were going to move to the market, in spite of the known issues.

They wanted to utilize the full plant capacity. The product was introduced in 2008 as ALWAYS Infinity "FlexFoam".- at a premium price.

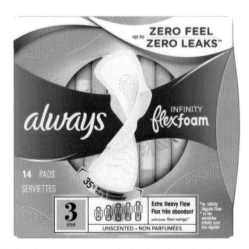

It didn't take long to see that the product was a "Game Changer" - an absolute new level of protection & comfort! The common consumer feedback was "IT WORKS LIKE MAGIC"

It took over 20 years from the first experiment, with over 60 patents generated, and great courage and brilliant work by many along the way. The same kind of path that has been travelled on most "Game Changers". A sound technology position has been achieved, and great progress with the technology will continue to be made. The outcome of all this tremendous innovative work is a breakthrough product and **a platform technology for the Company.**

One can only feel "The best is yet to come"

# SECTION FIVE

## "COMPANY STRUCTURES FOR MAJOR INNOVATION"

v

# 27.

# THE COMPANY INTERNAL VENTURE APPROACH

When I met with John Smale, after I became Chief Technology Officer in 1987, I remember his stern message.

**"Outside of geographical expansion, this company has only had significant growth from major new brands, or acquisitions. Growing with new brands is by far the best financial approach, and I'll be counting on you for that"**

That message never left me. While developing new brands is financially attractive, they are also the **toughest** challenge for any R&D organization. Not only for identifying the unobvious consumer insight, but for the very difficult technology development. All this assumes there is sufficient funding and adequate time for the development.

All companies struggle with the right structure to develop new brands. P&G had always been fortunate enough to have a very productive Corporate R&D organization. They initiated the development of many great products (see Chapter **26**). Further, we have always been able to develop new brands that were a new category within an existing business. Examples include Pampers Pull-ups, Rely & Always Feminine Protection in the Paper Category….Tide Liquid and Bounce in the

Laundry Category......Cascade auto dishwashing detergent in the Dishwashing Category, Pantene Shampoo & Conditioner and many others. But it was very unlikely that a major new brand in an entirely new category would emerge from our existing operating businesses. History is a great tutor here.

 Durk Jager, the P&G COO & PRESIDENT in 1995, and I had been discussing the need for new brands and our options. We might be able to charge a given operating business; or set up a new business unit to take on entirely new business development. But the experience with these approaches both internally and from other companies, was anything but positive. To be successful, there had to be freedom for discontinuous and disruptive thinking. There had to be isolation from the constraints of current business demands. We were reminded of the sage comments of Jay Galbraith: "an organization good at operating isn't good at innovating. They are opposing logics." When you then factored in the objective of many employees that valued fast job changes in their careers, we concluded that trying to do this in an existing business unit was unlikely to succeed.

At the time, we did have significant, major investment going into the development of a new osteoporosis drug in our small pharmaceutical business, because of our great expertise in calcium control (Chapter 19). But, we were searching for an approach that would increase our efforts in new brand development. We knew the Company was unmatched in the quality of its people. We had some of the brightest and most creative people of any organization in the world, and had that asset throughout the Company's history. But fully tapping the inherent creativity of the organization, and identifying and supporting the best ideas that might be available in a large Company is a very difficult challenge.

Among our employees, it was likely that a large number of great ideas existed for innovations different than what the people were focused on in their work assignments. Outside venture funding of new global ideas was becoming more prevalent at the time, particularly in Silicon Valley. The thought emerged - **What if we tried to seek these ideas broadly from our people, and set up an internal venture fund to move the best ideas forward?**

Durk and I discussed the Venture concept, massaged and built on the idea with Craig Wynett, who was running a small Corporate New Venture group, and evolved to giving it a try. Further, Durk wanted to be fully involved. I was thrilled with that. We established the program under a label of "THE INNOVATION LEADERSHIP TEAM" (ILT). Durk & I co-authored a letter to the entire global R&D organization. The idea was simple. We wanted to hear of new product ideas - BIG IDEAS - with the potential to create a $100 MM global product. We quickly moved that up to $500MM, and later to $1billion. It wasn't that these projections were expected be anywhere near precise….but we wanted to focus on big ideas, and see the vision of the proposer.

We put high emphasis on the total proposition…..the consumer concept…..the technology behind the product…and to hear the passion of the innovator. The proposer would be asked to meet with the "ILT" composed of Durk and I, John Pepper then CEO, and various business leaders depending on their availability - Bruce Byrnes, Wolfgang Berndt, Gary Martin, along with CFO Clayt Daley.

We put support systems and some structural changes in place to support the process. A small marketing group was established to help people develop their vision. An open-job bidding process was created, to provide staffing for the projects that were funded to Proof-of Concept. Most importantly we gave the individual or research team TIME and a small budget. They were allowed full time to develop the idea or rough prototype. None of this "do it in your spare time or on weekends". We stressed that there never would be personal career penalties involved.

The funding process then evolved to reducing each existing category's budget, across-the-board by about 10%. This created a large enough fund to fully get things underway.

A complete change in our patent licensing policy opened all technologies up for licensing and potential donations. Also, critically important was the reallocation of R&D investments from operations that were judged to have weak programs into the Venture Fund.

I don't think that any one of us realized just how many creative product ideas existed in the reservoir of researchers' minds. We were excited, because as Linus Pauling, the two-time Nobel Prize winner said,

## "The best way to have a good idea is to have lots of ideas"

The new ideas that came forward were quite amazing. The enthusiasm and pride of individuals getting the chance to talk about their ideas with the leadership of the Company was energizing to the presenters.

Word spread at internet speed to all parts of the Company. For the projects that received funding, the ILT members pressed the recipient to establish the next milestone. What needed to be achieved in order to find out whether the project had a chance to work? Success meant a new round of funding. It wasn't exactly like Silicon Valley venture capital, but it came pretty close.

Many entirely new products got to the market and were successful: **(CASE STUDIES ON SEVERAL WILL FOLLOW):**

SWIFFER DUSTING PRODUCTS & SWIFFER WET JET MOPS - global, MULTI-BILLION DOLLAR new brand.

FEBREZE - global MUILTI-BILLION DOLLAR odor-capturing new brand

GLAD PRESS & SEAL WRAP - GLAD INCREASED STRENGTH STORAGE BAGS & ODOR CONTROL - true breakthroughs in home plastic products - licensed to Clorox & broadly marketed as GLAD products.

THERMACARE HEAT-GENERATING PAIN CONTROL WRAPS - national, sold to Wyeth/Pfizer.

CREST WHITE STRIPS - new, whitening breakthrough product that became a global brand almost instantaneously - Idea originated from the Technology Council (Chapter **29**)

OLAY FACIALS - new product segment in Olay Skin Care Line globally. Provides cleanser, makeup remover, toner, scrub and mask in one. Idea originated from the Technology Council (Chapter **29**)

ALIGN PROBIOTIC - The idea of using a probiotic or the ingestion of a desirable bacterium, to improve gut health, was gaining the support of leading gastroenteroligists. The product was the first to use Bifodobacterium 35624 and had unique packaging to maintain stability. It was the first probiotic to demonstrate positive effects on Irritable Bowel Syndrome. The brand was named ALIGN, and nationally expanded in the US. Unfortunately, the full envisioned potential was never realized, at least until today.

WRINKLE-REDUCING SPRAY - now sold as DOWNY WRINKLE GUARD nationally

FIT NATURAL FRUIT & VEGETABLE WASH - patented all-natural formula, test marketed, sold to HealthPro Brands, and is sold globally

DRYEL DRYER-ADDED DRY-CLEANING/FRESHENING NEW PRODUCT - patented approach to do dry cleaning in a clothes dryer, test-marketed, and sold to Summit Brands

TIDE LAUNDRY & CLEANER STORES - evolved to 125 franchised stores today in 22 states

MR. CLEAN CAR WASH - evolved to 18 franchised car wash operations today

"SWATCH" HOME DRY-CLEANING - development of a home dry-cleaning machine with Whirlpool, which became a Whirlpool owned and branded product...perhaps ahead of its time.

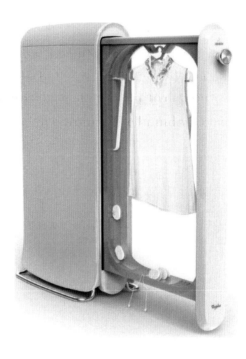

Obviously, not every project worked. There was a breakthrough hand sanitizing formula that got incorporated into a commercial public restroom system..but an unexpected safety issue emerged. A breakthrough was envisioned for a laundry product uniquely designed for washing athletic/everyday shoes. These shoes at the time were almost all white and leather, and a huge problem for families to keep clean. Safely cleaning leather was a big deal....but almost overnight, these shoes broadly in the market went from leather to fabric structures, and the product idea was no longer relevant.

Overall, the program was highly productive....but the main business advances didn't happen until some 3- 5 years following the venture product launches. In 2000, I had planned my retirement after 13 years as CTO, and unexpectedly Durk Jager voluntarily retired following a significant profit shortfall. He had been aggressively pushing a global organization restructure and broad cost reduction, with all the inherent challenges, and normal operations were affected.

That global business structure, however, became a pivotal leading-edge change for P&G moving forward through today.

The unexpected departure of Durk Jager, along with my retirement at the same time left the venture program vulnerable, and viewed as an overly risky, unacceptable "corporate tax". Further, because of the profit shortfall, effort had shifted to focus on the existing brands. When the benefits of the venture program came to be realized several years later, and recognized publicly by shareholders, the leadership interest was no longer in place - or the importance understood -and the Internal Venture Program was never re-instated in the original format.

Whether this specific structure is still right or would work today is highly problematic. I do know that the ideas, energy, passion, and excitement that emerged around the innovation needs of the Company were at an all-time high level in my 40 years of experience. The elements that were working at that time are probably as right today as before.

# NEW BRAND
# SWIFFER SWEEPER,
# WET-JET MOP, & DUSTERS

P&G global household cleaning brands in 1995, consisted of Mr. Clean, Mr. Proper,, Spic & Span, & Flash multi-purpose cleaners & Comet cleanser. The products were introduced in the 1950's, represented a small volume and low-profit category, and in the case of floor cleaning, the consumer hated the process. Clearly, this was difficult to bear for a company with a rich history of selling solutions to consumer problems.

 US researcher, Paul Russo, and his director, Wilbur Strickland, had undertaken a "skunk-works" project on floor-cleaning. Paul had already clearly proven that the choice of mop type by the consumer

was FAR more important than the cleaner product for achieving the best end result. He had been experimenting with different potentially one-use, disposable pads as a new type of mop. However, the idea that was emerging with Paul and Wilber had not yet been exposed to their business partners, or made an official development project.

At about the same time, the KAO company in Japan introduced a "sweeper" product with replaceable nonwoven sheets for daily dust removable on the household tatami mats present in every Japanese home. While a very interesting product, it was a niche product focused on a narrow task.

Paul & Wilbur gathered their thoughts for what could be, and with the idea of a floor "sweeper" and a "wet-pad mop", targeted a session with the venture-supporting Innovation Leadership Team. It didn't take long in that meeting to see that this was clearly a breakthrough idea, and with our competencies in cleaning chemistries, paper, and absorbent pads….it could be a big winner for P&G. It didn't take Durk Jager long to utter…"Let's get going"

Claude Mancel, the French national, global R&D head for Home Care, and for cleaning products, asked…What about funding? I indicated…shut everything else down, and put everything in your budget against this…and after some aggressive "push-back"… off we went.

It was a **huge** challenge. We were into multiple implements, where we didn't have experience. We had KAO patents to contend with. There were critical design needs for the various parts. We had multiple complex technologies to bring together, framed in a strong sense of urgency.

 The winning strategy from the start was to form a strong, unified team of R&D and Product Supply. Bob Bradford & Michael Sullivan from Product Supply were "joined at the hip" with Strickland and Russo. They were able to attack design, suppliers, logistics, and manufacturing loca-  tions as one of the most unified teams ever formed at P&G. Bradford & Sullivan also had their official assigned responsibilities for several P&G plants to manage, as they also took on the added challenges of the SWIFFER development. So it was clearly "the call to duty" that drove their effort.

The R&D team was able to bring in a few very experienced people from Corporate Packaging for plastic molding, from Paper for non-woven fabric design, and from Diapers for absorbent core design. These key people were critical to bring the needed technologies to bear on the project.

The Kenner Company, a major toy company that was headquartered in Cincinnati, had designed the enormously successful "Star Wars" toys. They were contacted for design guidance on handling multiple integrated plastic parts. They put the team in touch with their former designers. The experienced designers provided outstanding help in educating the team on the "mechanics and pitfalls" of the process, and how to bring the plastic parts to the market at scale, with good financials…and speed. The several KAO patents on implement design features were circumvented with new patented designs that yielded performance and convenience improvements.

After extensive searching, China suppliers were found to have the most advanced plastic molding technology by far, and they were engaged.

The nonwoven supplier, PGI (now Berry Plastics) with their operation in Karwijk, Holland was the development supplier for the sweeper sheet. KAO'S sheet was designed for only light dust pickup off a tatami mat. The team had to obtain a sheet that could handle larger dirt particle pickup. Using P&G's unique structured paper technology applied to a nonwoven fabric production machine, a unique "micro-3D" structure was constructed that met the goal.

The absorbent core was also a challenge. With diapers, liquid naturally enters the core. With SWIFFER, the core needs to **draw** the liquid in… working against gravity. A unique core was created with a surfactant to aid the water pick-up.

So….plastic parts from China…sweeper sheet from Holland… cleaner bottles and absorbent cores from P&G plants…all brought together in a Green Bay, Wisconsin assembly plant. A logistic nightmare…but wonderfully handled.

Test markets of the SWIFFER SWEEPER were started in France and the US in 1998…and the brand was an immediate success. Global expansion started a year later. Test markets of the SWIFFER WET-JET were started in 2000, and expanded globally in 2001.

A patented "tow fiber" was licensed from Unicharm in Japan - a KAO competitor - and it formed the basis for a line of SWIFFER DUSTERS, which were launched in late 2002. This extended the vision which had been established early that the SWIFFER brand could be a family of better and different household cleaning products - and will likely see the continued expansion of new products into the future.

# The **SWIFFER** brand reached $1billion sales the fastest of any P&G brand in history.

# NEW BRAND
# FEBREZE

Air fresheners were first introduced in the mid 1950s, and the common technical approach was to use fragrances to mask the unwanted odor. Every conceivable form of distributing fragrance in spaces evolved over the future years. Whether it was candles, aerosols, sprays, impregnated fabrics, electrical plug-in devices that vaporized fragrances, automobile vent attachments, and many others. There were also more dedicated machine devices with a variety of technical approaches - none of which was highly effective. For example, one machine moved the room air over activated charcoal, creating ozone, which really was found not to block the odors, but to block the human ability to sense the odors, which could be harmful at high levels.

Air freshening was a market that P&G never developed interest in. The technology was very ordinary, and amenable to multiple competitors entering the market based upon very small product differences.

Toan Trinh was a P&G researcher who came to the US from Vietnam. Under Japanese and North Vietnamese interventions, he had spent the first 30 years of his life in a war zone. He was a brilliant scientist, who had done his doctoral work on inorganic chemistry and chemical engineering at the University of Wisconsin. Subsequently, he taught at Saigon University. He was an extremely well-liked researcher, with unbridled energy, and a constant smile, which always set up a penetrating question, or a unique perspective. He categorized his approach to innovating as a "3P" style - Persistence, Persistence, and Persistence! Among his experiences, he had spent considerable time working with a chemical compound called cyclodextrin.

Cyclodextrin is a complex sugar molecule. To a non-expert chemist, the chemical structure would look like a unique circular "cage". Cyclodextrin has many different chemical properties, and is used in the pharmaceutical industry for increasing solubility, bioavailability and stability of drug compounds.

Toan was passionate about the ability of Cyclodextrin compounds to encapsulate malodors. Because of their unusual structure they could actually trap and retain water-soluble compounds. He arranged a session with the Innovation Leadership Team to bring his idea for a new air freshener forward.

His concept would be the first product that would actually **trap odors** as opposed to masking them. It would be effective, long lasting, safe, and have a myriad of applications for all situations where malodor was a problem. Further, basic patents he had filed were already issued and the technology offered the opportunity for many additional patents.

He brought in a large number of demonstrations showing the odor-trapping effect of the cyclodextrin technology. They were impressive. We spent much of the meeting experimenting ourselves with the different delivery devices he brought in to demonstrate.

He painted a picture of applications of the cyclodextrin in almost every kind of consumer product imaginable. There would be direct applications in aerosols, sprays and plug-ins. The technology could also be used in many P&G products like laundry detergents, fabric softeners, cleaners and countless others.

# He painted a picture of a multi-billion-dollar brand!

There obviously was a great deal of skepticism to much of what he presented because the effectiveness seemed a bit difficult to fully accept. But there was no way one couldn't see the tremendous passion he had for the idea. It was contagious and very exciting.

The project was accepted for advancement. A team in the Global Laundry & Cleaning Products organization was created to develop prototype products and get them exposed to consumers. This proceeded well and initial reactions from consumers were very positive. The product was used in typical air freshening situations for room deodorizing and the product was rated better than the leading brand at the time. The marketing organization was relatively cool to the idea, because of the relatively low margins and fragmented category that existed for air fresheners. But they agreed that the concept of trapping rather than masking odors was a very powerful one with consumers - if it could be achieved.

A plan was put together to conduct a market test. A brand name was chosen – FEBREZE,-, The product was going to be sold in an aerosol and also a spray bottle for fabric applications. Package artwork was put together and the new brand was ready to go.

It was assumed that this product would be appreciated particularly by high-end consumers, who were looking for the ultimate in dealing with malodor in their homes.

The market test was started. In just a short period of time we were blown away by the tremendous positive reaction we were getting from consumers. We were not only seeing the expected use of the product for room air freshening, but unexpectedly it was strongly supported by relatively modest and low income families. Older household furnishings like sofas, chairs, drapes, and rugs, which had seen the best of times, had developed malodors which were not easily eliminated. Suddenly they took on a new level of freshness with the application of FEBREZE. **Consumers were not positive, they were ecstatic**! College students used Febreze as an easy and affordable way to avoid laundering, and a new way to handle a bad Saturday night out. We were really stunned with the amazing positive consumer reaction to the product, and the amazing breadth of applications.

Plans were readied for national expansion, and exposure to international markets. International consumers on a very broad basis were as excited for the product as what we had seen in the United States. Febreze appeared to be a breakaway major new global Company brand.

Over time, the applications that Toan had forecasted all seemed to come true. Febreeze was used as a great additive to laundry brands and our fabric softener brands. It eventually worked into Glad garbage bags, and a host of other products. Over time, the brand name FEBREZE became the worldwide standard for malodor control - an iconic brand.

A marvelous multi-billion-dollar new brand brought about by the tremendous passion and knowledge of a **single inventor...** who got the support to "give it a try". An absolutely great invention, ranking right at the top with other iconic Procter & Gamble innovations.

# -MAJOR PRODUCT ADVANCES-
# (GLAD) PRESS & SEAL WRAP,
# and "FORCEFLEX" & "ODOR
# PROTECTION" TRASH BAGS.

P&G has a highly competent, state-of-the-art, corporate packaging and prototyping group that brings expertise to the entire stable of global product categories.

> *With the advent of the Company's New Venture program George Vernon, the director of the group and his packaging researchers, brought forward ideas for major new products based on their deep expertise. Of particular high-level interest were ideas that could bring major new benefits to the consumer's line-up of "cling" household wraps and also trash bags.*

> *"Cling" wraps were ubiquitous in household kitchens. Made from low-density Polyethylene or Polyvinyl Chloride (PVC), these normally clear films, tailored for*

*some natural stretchability and stickiness, had "every-day" use for a great variety of sealing applications.*

*If there was a main consumer need in the existing products, it was that existing products did not adhere nor seal well to plastic and wood surfaces. Many household storage containers were made from these materials, and consumers were frustrated with seals that didn't adhere or broke, and having to transfer food or other items to glass or ceramic containers before sealing and storing.*

*At the same time, work was going on in the Oral Care Division applying peroxide gel to plastic film for the development of CREST WHITE STRIPS. The idea evolved to apply an adhesive to the film to create a wrap that would provide good sealing for every application, including plastic and wood, and also include the ability to form a" pouch" by sealing the wrap to itself. The film surface was altered with "micro-dimples" to hold the adhesive. Prototypes were prepared and tested, and consumers really liked the idea. Manufacturing know-how for applying glue to the film existed from similar experience with PAMPERS fabrication. So, feasibility seemed very doable.*

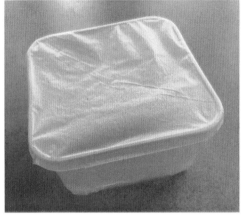

The precise properties of the adhesive were critical. A dry surface was needed so the wrap could be rolled and placed in a package, yet offered good adherence when pressed against a surface. (a la' Post-it notes). The adhesive experts went to work, and 'PRESS'N SEAL PLASTIC WRAP was the result!

Using the application of other P&G technologies, new highly desirable benefits were also envisioned for common trash bags. There was "ring-rolling" proprietary technology, again from Pampers, to create stretch in a plastic 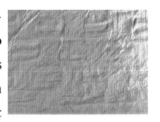 film. It involved creating continuous micro-creases in the trash bag film. When the film was subject to a force it would have some stretch without breaking. Voila!, "FORCEFLEX TRASH BAGS" were created!

P&G also had immense fragrance experience and, within their capabilities, was a great knowledge of low-volatility, highly stable long-lasting consumer preferred fragrances. Surprising to the researchers was the fact that providing a pleasant "odor control material" to trash bags hadn't really ever been pursued in the market.

Apparently, trash was supposed to demand needing to be "taken out". But consumers didn't need much persuasion to quickly see the great benefits of masking trash odor. Steady release "ODOR PROTECTION

TRASH BAGS" were created, and consumers loved the product! Why wouldn't they!

So a strong line-up of exciting technologies were available for brand new consumer-preferred products. A string of strong patents were filed and granted on the technologies.

Were these products enough to launch an entirely new brand in the plastic wrap/bag category? Clorox (Glad) and S.C. Johnson (Ziploc & Saran) were leaders in the US, with a full line of products, established manufacturing facilities and proprietary bag closure technology. It was going to be tough; they were going to be formidable competitors. Although the decision was made after I retired, the options were thoroughly evaluated and it was decided to license the P&G technologies to Clorox. Clorox launched "Press 'n Seal food wrap, and the "FORCEFLEX" and "ODOR PROTECTION" benefits in their GLAD trash bags.

All were very successful new-to-the-world additions to the GLAD product line-up. When FEBREZE odor capture technology became a P&G reality, it provided yet another technology that was incorporated into GLAD trash bags, and which provided Clorox with a major competitive advantage with an iconic odor protection asset.. All of these applications are great examples of connecting and using Company technologies across business units, AND the effectiveness of the Company's New Venture program.

# As a result of these great innovations, Clorox licensing royalties became an ongoing profit contributor to P&G's bottom line.

## NEW BRAND
## THERMACARE DISPOSABLE HEAT-ACTIVATED PAIN RELIEF PADS

The P&G New Venture structure had a small marketing group - Corporate New Ventures (CNV) - as a "broker" for new ideas as they came from within the entire Company. One particular upcoming ILT meeting had a discussion of a Disposable Heat-Activated Pain Relief product on the agenda. It certainly looked very interesting. Craig Wynett, the manager of the CNV group, and Charles Hong, an experienced entrepreneurial researcher were scheduled to present the idea. We all were looking forward to the discussion.

The product design was made up of a non-woven sheet impregnated with popular pain rub actives (camphor, capsicum, menthol & methyl salicylate)…along with pockets in the sheet that contained a unique compound. The heat-generating approach - which was very surprising to all of us, and particularly me - was going to be controlled iron rusting. Sounded a little crazy at first blush. Although new to most of us, the "technology" had been used for some time in disposable hand warmers.

Finely-ground, moistened iron particles were contained in the sheet pouches. Activated charcoal was used in combination with the iron particles to distribute the moisture evenly . The entire sheet was enclosed in an air-tight outer envelope. When the sheet was removed from the envelope, oxygen in the air would start the iron oxidation, or rusting process, creating heat. The combination of the right size and correct amount of iron particles in the pouches and the right porosity of the non-woven sheet provided air access. This would be the mechanism to control the temperature and duration of the heat.

Samples had been hand prepared, and we all had a chance to try them. By gosh, they worked! The total sensory experience of providing heat and pain relief was very unique and positive.

The entire concept provided a great opportunity for a proprietary position given the well known benefits of applying heat to ameliorate the pain and discomfort of a sore muscle, joint and tissue. We all felt that the product could clearly be a superior way to provide topical pain relief.

Controlling temperature in a hand warmer provides flexibility, and any chance of a "burn" is extremely low. The Thermacare prototype presented huge technical challenges considering the number of different parts of the body where it might be applied, and the attendant performance, safety concerns, and wearability comfort challenges that needed to be addressed.

Joan Szkutak was the director in the OTC Health Care R&D organization. She faced a major challenge in acquiring the diverse talent needed to tackle this highly challenging, and entirely new technology-based product.

Ron Cramer, an experienced team leader in her organization was available for this challenging assignment. A full contingent of Pampers experts jumped on board excited to use their experience to tackle this unique challenge. Bill Oullette, materials, Janice Urbanik (process), Leane Davis (product design), and Carl Noble (engineering) quickly formed the team to create a practical product that could be manufactured. They also discovered that one of our research physicians had expertise in pain physiology and treatment. He provided very critical consultation to the group.

If you enjoyed advanced thermodynamics in college then you wanted to be on this development team. The challenge was similar to a nuclear reactor where the amount of heat generated had to be precisely controlled - in this case for 8-12 hours. A sufficient amount had to be generated to conduct through the pad and a layer of fabric to provide a steady amount of heat without a surge or abrupt drop off.

The product promise was to be "**Wearable Heat for Safe Pain Relief**". A variety of different designs for back, neck, shoulder, menstrual, knee & multi-purpose had to be designed and prototyped. THERMACARE couldn't be 8 different designs of the basic heating pad and have an affordable product . The goal was to have one basic pad design, and then change the shape of the pad for each product. The goal was to create a "garment" for the application, not just a warming pad that got applied to the area, as some poor performing market products had done

There were, of course, medical concerns placing a warm pad on the body for as much as 8 hours. The device had to be comfortable to wear for a full day. The manufacturing challenges were immense.

The Corporate Prototyping Group (Chapter **28**) was a major contributor. They designed an apertured film which encased the iron actives, and created "heat cells". This way the oxygen flow was consistently and carefully controlled by the film, as opposed to depending on the porosity of the fabric cover. They also created a small scale pilot plant to make the heat cells, as well as finished product.

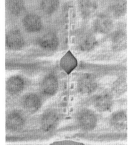

Over the next several months the team created a standardized pad design that had the correct amount of active materials. It could be flexibly converted to each shape while providing wearability comfort, and could be manufactured efficiently.

A test market production line was designed and installed, and the test was launched. Altogether, the variety of shapes and sizes created a very substantial and attractive line-up of products and store display space for the launch.

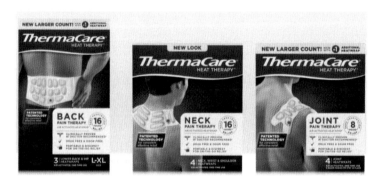

Numerous patents were ultimately issued on all the features of the pad's design and composition. Fortunately, producing the uniquely designed pads fit nicely into P&G's broad experience with Pampers, Always, Cascade and Tide Pods, and Bounce brands, which yielded excellent manufacturing execution.

The product launch was successful, and THERMACARE became a broadly available new brand for pain relief in US stores. A number of clinical studies showing Thermacare's performance advantages were published shortly after its launch. One study in particular done at the Rutgers Medical School showed THERMACARE performed better than either Ibuprofen or Acetaminophen for acute lower back pain. THERMACARE gained support from doctors as well as consumers.

Six years after its launch the brand had reached reported global sales of $100 MM . Not quite large enough for a place in P&G's portfolio. During a time of brand consolidation THERMACARE was sold to Wyeth Pharmaceuticals, who also marketed Advil (Ibuprofen) at an undisclosed amount.

THERMACARE is a clear example of the great creativity, technical competence and determination that can be harnessed when an exciting consumer goal is used to challenge a group of highly capable researchers. Particularly when they are given the freedom and support to accomplish it.

It took the combined efforts of experience in PAMPERS, Corporate Prototyping, and Health Care. It probably never would have happened if one of the organizations had considered doing it by themselves. **Certainly a GREAT example of leveraging a Company's existing technology strengths and people assets**.

# 28.

# THE EVOLVING ROLES OF BUSINESS-FOCUSED & CORPORATE R&D

With the long history of Procter & Gamble, one can follow the evolving structure of R&D within the company as it grew.

In the early years there was a small group of chemists whose work would best be described as factory service and quality control. Once IVORY was developed there was a need for chemical analysts to support Ivory's claim of 99 44/00% pure. So P&G's strong history of analytical chemistry goes back to the founding years.

There really wasn't much corporate R&D in the United States before the beginning of the 20th century. It was mostly individual inventors/entrepreneurs like Thomas Edison and Nikola Tesla. The first real research labs were developed at AT&T, General Electric - who had inherited Edison's Menlo Park, New Jersey lab - Westinghouse (Tesla) and Kodak. DuPont research began with the hiring of Wallace Carothers, inventor of nylon in1928. But the US was a backwater to Germany and Great Britain where basic work on telephony, aniline dyes, electric lighting, long-wave transmission and ammonia refrigeration was being developed. World War I changed all of this.

Real research at P&G probably began about the time Victor Mills was hired in 1927. There was also a group of engineers hired at the same time who enjoyed 40+-year careers at P&G and created the ethos of research and product development. This included engineers like Bill Martin and Jud Sanders, who became key managers in Food R&D, where Bill and I started.

As the company grew, this "Corporate R&D" group freely explored opportunities to build off its technical strengths and also new technologies acquired from acquisitions. It remained tied into the Company's top management.

In the early 1950's, business groups needed more effort focused on improving existing brands, and business product development groups were started. By 1955, 80% of the R&D effort was corporate, and 20% tied to the business groups.

At this time the Corporate R&D group was amazingly effective, with the ability of researchers to work closely together on cross-fertilization of technologies and ideas. **They could connect technologies around the lunch table**. Paper, Coffee & Food acquisitions brought in new technical capabilities. Pampers, Pringles, Crisco, Crisco Oil, cake mix processing, coffee and nut roasting technologies, continuous soap hydrolyzing, spray drying advancements, and many other business producing inventions were conceptualized and put on paths to commercialization. Communication of the R&D work with Company decision makers was straightforward, and innovation moved to the marketplace.

All the key elements for optimum innovation were present....**CLOSE CONNECTION OF DIVERSE TECHNOLOGIES, FREEDOM, and A DIRECT TIE WITH TOP COMPANY MANAGEMENT.....**and, of course, the presence of **outstanding innovators.**

As time went on, more of the R&D effort was put into the business units to support the continual need to upgrade the expanding number of existing brands, and to support international expansion. By 1990, Corporate R&D was reduced to 25%, the US was at 45%, and International 30% of the total R&D staffing. The Corporate work was separated into groups supporting upstream technology for each of the major operating divisions. There were several examples that showed that this approach was effective, The prime example was hair care, where Franco Spadini leading a corporate R&D group and Ceil Kuzma's Beauty Care Development group, collaborated to create a shampoo & conditioner in-one, and the major global brand, PANTENE. (Chapter **26**)

In the late 1990's, as P&G moved to a global structure, the corporate R&D groups were incorporated into the global business structure, reporting to the global business sector R&D leader. The challenge here is for the global R&D leader for a given business to maintain upstream, leading edge innovative work, with the pressure of their boss wanting more work on the current business. This is the conundrum of the innovation process and the creative destruction that has to be programmed. Not for the faint of heart. Safi Bahcall in his book "Loonshots", and the study of past major innovations, said, "People responsible for developing high-risk, early stage ideas need to be sheltered from the soldiers responsible for already-successful steady-growth part of the organization."

In this scenario, the CTO plays a significant role in influencing the right balance in the portfolio of work. Critical overall is to keep one's eye on the current business, but also keep work active for the big ideas, and game-changers longer term.

SO….in a large, multi-category global business, what is the role of Corporate R&D, if any?

There clearly is a need for core capabilities applicable to ALL operations:

CORE PACKAGING & PROTOTYPING - This is the equipment and people who keep the company at the leading edge of the fast-moving technology that is important to all businesses.

This provides a **critical** capability. The group provides the leading edge packaging, engineering, and product prototype design expertise to bring an idea to life. At P&G, this group's mission was "If you can dream it, we can make it."

When you consider the prototype needs for products I've discussed like SWIFFER, THERMACARE, CREST WHITE STRIPS, FORCEFLEX TRASH BAGS, PRESS & SEAL WRAP, and many, many more, you can easily understand the difficulty in 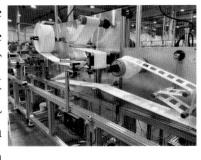 getting feasible products constructed from a basic idea. Flexible pilot plant equipment to do almost any kind of manufacturing function was available, along with experienced researchers who could quickly move to producing product samples. Initially hand made samples, and then to pilot plant capability to produce product for consumer testing.

One of the critical factors in bringing an innovation successfully to the marketplace, is the partnership that can be fostered between R&D, Engineering, and Product Supply. History is rampant with troubled projects where this group could not form an effective partnership, and projects lost valuable time, had significant inefficiencies, unnecessary costs, and even were responsible for failed projects. The key for improvement was felt to be early involvement of all the parties….. working together from the start. The structure that evolved at P&G was the creation of a side-by-side development center where R&D could develop the prototype and package concept, and connect from the start with the other functions.

An experienced engineering group in a technical center right next door to the prototyping center work hand-in-hand to not only get a feasible process, but one that builds in desirable scale-up design. Further, the entire diverse range of P&G product technology experience is represented, so almost every new product technology need could be met. A truly wonderful example of connecting all the Company's existing product design capability, as well as new leading-edge manufacturing design technology in one center. The speed and quality of product creation capability was amazing to me, and critical to project success

## HUMAN SAFETY, ENVIRONMENTAL & REGULATORY SERVICES

These are absolutely critical skills for bringing innovative products to global markets, and they can best be centralized corporately. This allows building deep expertise, and avoiding duplication across businesses as well as excessive costs. At P&G, they are organized in a group termed

**GLOBAL PRODUCT STEWARDSHIP.**

With over 5 billion exposures of P&G products a day, it's a huge challenge to insure safety to consumers and the environment, compliance with all regulations, and alignment with P&G's sustainability vision.

With core competencies in toxicology and environmental science, along with deep expertise in a broad range of global regulatory pathways involving some 60 countries, the Product Stewardship group works hand-in-hand with the product development teams to create the desired innovative products.

In toxicology, P&G invested more than $500MM in animal alternative test approaches and after decades of work **eliminated the need for animal tests, except when required by government regulation in very few situations (e.g. China).** Every toxicology pathway identified from the decades of work with animal testing, as well as human experience, was transformed into a computational toxicology platform. This achievement was thought impossible for decades. The outcome was an amazing innovation! P&G led the world in almost every area, but particularly in understanding skin and respiratory sensitization - critical for safe use of key ingredients like detergent enzymes and perfumes. The achievement was recognized broadly within the scientific community, and also applauded by the large group of consumers urging elimination of animal testing. Internally, two researchers leading this work were appointed to the Victor Mills Honorary Society: Frank Gerberick and George Daston

In the environmental area, P&G developed unique laboratory capabilities for biodegradability and life cycle assessments. P&G for decades led the world's understanding of biodegradability on key ingredients, like surfactants, softeners, detergent builders, and polymers - pioneering scientific techniques, including the highly unique use of an actual  river stream laboratory. Leading the environmental scientific work was Robert Larson, whose work was acclaimed by global peers, and also through election into the inaugural Victor Mills Honorary Society.

Carbon footprints for all P&G's major brands have been developed, pointing the way for innovation opportunities producing environmental benefits. For example, laundry wash temperature was identified as the biggest opportunity for detergents, spurring increased technology focus on effective cold water washing.

With P&G's extremely diverse range of product categories and global markets, it has the broadest range of regulatory pathways of any consumer goods company. With current societal expectations and public policies, the number of chemical and product regulations continue to increase. Given this situation, credible science and data are critical to guide the formation of regulations that continue to support innovation.

But, then should there also be:

## CTO DIRECTED WORK?...

...**I believe this is very important**. It can be platform process, packaging, or ingredient work that has broad application in the company. It could be the core of a venture program with teams working to create entirely new brands, or new product segments - and having direct interfacing with the top management of the company. It could be a particular development partnership with another supplier or manufacturer to create a new major product.

Looking across companies today, it appears more frequently that this effort is being eliminated. From the experience P&G has shown over its entire history, this is a major error. The **FREEDOM AND DIRECT TIE WITH COMPANY DECISION MAKERS** are proven parameters for effective innovation.

# SECTION SIX

## "FOSTERING CONNECTIONS FOR INNOVATION"

# 29.

## CREATING THE GLOBAL "LUNCH TABLE"

When I began my career, cross-technology and cross-business connections could be done over a "lunch-table". Of course, as the Company grows and becomes more decentralized - eventually global- this becomes a huge challenge To help recreate that "lunch-table", a Global Technology Council was created which, as CTO, I chaired. It was made up of our business unit R&D directors, corporate R&D lab heads, and key geographical R&D Managers. (SEE CHART OF TECH COUNCIL WHICH FOLLOWS) Most of these individuals had a dual reporting relationship to me - and to a business unit. It was designed to bring together knowledge from each of our different business categories, our different major geographical areas, and the corporate group.

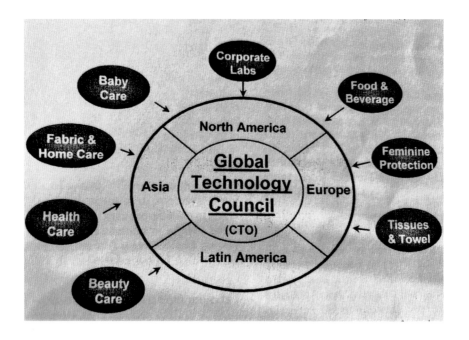

This working forum was designed to leverage our existing technologies and capabilities **TO SOLVE IMPORTANT PROBLEMS,** or at a minimum come up with specific new courses of action to be tried. A particular problem or opportunity was put on the agenda.....and everyone came ready to engage.

The interaction always seemed to generate entirely new ideas, and we didn't break up until we had at least one solid idea to pursue.

One example was to identify new applications for our sucrose-ester technology OLESTRA. We knew we had a unique and highly proprietary technology. The material had very unique physical properties. We initiated exploratory product work using a small scouting group. One idea quickly connected. The physical properties of OLESTRA permitted moisturizing benefits to the skin in a cleansing product. It also prevented moisture loss from the skin after use without being greasy to the skin. It became central to a new product concept - a convenient daily facial cloth, impregnated with ingredients, that when wet, would lather, and provide five benefits in one. A cleaner, makeup remover, toner, scrub and mask in one easy to use and easy to dispose product. The key substrate, and ingredients to be explored were all identified. The concept was presented to the Innovation Leadership Team, accepted, developed, and expanded broadly as OLAY DAILY FACIALS. We might never would have gotten to this application without a conscious effort to explore unobvious connections with a group who could make the connections, and force a try.

Another very key example was teeth whitening. Barbara Slatt was the US health care R&D director. At a meeting, she was asked to provide an overview of existing ways for consumers to achieve advanced whitening of their teeth. Performance from specialty pastes and brushing  was very modest, and disappointed most users. The common way for significant whitening was with a whitening kit….a combination of peroxide bleach and a mold of the actual teeth to encase the teeth. The device was purchased from a small entrepreneurial OTC marketer or a dentist. The device was a formed plastic or rubber implement that needed to maintain close contact of the bleach with the teeth. The device needed to be worn for extended periods of time, …usually overnight for a week or more. The consumer was held hostage using the device, and had to bear the highly uncomfortable process in order to get a positive result. The devices that were not custom fitted rarely fit properly. Custom fitting a device with a dentist was time consuming and expensive - several hundred dollars to get into the game.

In spite of the unavailability of convenient, affordable, and effective products, consumers remained VERY interested in achieving better teeth whitening - a high level consumer need.

In the Technology Council discussion, ideas rattled around ways to better produce the existing high-end device products with greater convenience and lower costs.....but then, it shifted to "band-aid" applications. Pre-prepared strips of peroxide gel on a strip that the consumer applied. The manager of our corporate prototyping lab volunteered to create several prototypes for evaluation.. A polymeric film that was dimpled and filled with adhesive for food wraps was redesigned to hold a peroxide gel. BINGO! CREST WHITE STRIPS was created. Consumers were delighted.

To illustrate consumer interest in teeth whitening in specific situations, data indicated that in a particular wedding weekend, the entire wedding group would spend an average of $1500 on CREST WHITE STRIPS. ...Pretty amazing consumer interest!.... A digital KODAK moment!!

The Technology Council meetings were meetings I definitely looked forward to.. Challenging, Fun, Productive, and Relationship-Building. Not much lunch was consumed.

# 30.

# OPEN INNOVATION

In the discussion in Chapter **19**, "The Importance of Building From Your Technical Strengths" the point was made that, in spite of their critical importance, one cannot depend only on internal technology connections for innovation in today's world. The need to go to an OPEN INNOVATION Culture was absolutely necessary. Research & Development had to develop a significant component of "Connect & Develop" capability.

**A great deal of thought** went into how to create a major intervention to launch OPEN INNOVATION into the R&D organization. We had to strategize how best to engage key constituents. We had the challenge first with our own employees. Open Innovation represented a huge mindset change to our employees who operate on a "need to know" basis. Then we had to change our approach with suppliers whom we had traditionally worked over to get the lowest possible prices. We already had some links to universities but came nowhere near fully understanding their technology cache and willingness to collaborate. Same was true of the major Government Labs, like Sandia and Los Alamos, who were charged with marketing their non-proprietary technology.

We decided to take a VERY BOLD step. A **one-of-a-kind** "deal-making technology trading expo" was planned and launched as "INNOVATION 2000"

Everyone, sometime in their schooling, whether in grade school art class, or all the way through graduate school thesis presentations, has been involved in or observed a "poster session". A venue where everyone presented their work in some combination of visuals and real-time demonstrations, and was present to explain their work. Visitors could walk around and observe, and interact with the presenter.

# I suspect everyone's experience with such a "poster session" was to come away with at least one new learning or memorable thought.

Procter & Gamble executed a global R&D organization's version of this type of "Poster Session" with INNOVATION 2000.

This three-day exposition in June, 2000 showcased over 100 of P&G's most promising, cutting-edge technologies with a global audience consisting of R&D, Engineering, Marketing, and General Management - across all the Company's business sectors.

It was held at the spacious Cincinnati Convention Center, just down the street from P&G's downtown headquarters. Over 5,000 P&G researcher visits happened. For those who could not attend, the latest in webcasting and satellite technology was used to create our own Innovation News Network, INN, complete with news anchors, reporters and even commercials. Live broadcasts were done from the Convention Center, so that R&D employees from around the world could get access directly at their desktop computers. Also, hundreds of cell phones were distributed to P&G participants with the sole purpose of recording new ideas and new connections as they were discovered on the exhibit floor. All ideas were recorded into a central database.

The event received significant local media, and some national media coverage.

Taking this whole "connection-making expo" event a step further, we invited external suppliers to showcase their technologies as well. In addition to the P&G-sponsored technologies, there were over 600 representatives from about 50 exhibitors of non-P&G technologies - all under one roof at the Cincinnati Convention Center, right down the street from our world headquarters. Participants included developmental suppliers, university collaborators, federal laboratories, and research institutes from around the world. Japanese and European representatives even made the trip at their expense to participate in this showcase innovation event. This was the largest exposition of cutting edge technology we had ever assembled! It offered not just a glimmer of ideas in the minds of upstream, white-coat researchers, but technologies that had been developed to the stage of commercialization.

The partnerships with key suppliers was a clear accomplishment. Historically, P&G's approach was to solicit bids on the $14 billion in chemical and packaging materials used each year. Then the Company's significant purchasing power was used to drive down prices among competing suppliers. An alternate concept arose to develop "Critical

Supplier Partnerships". Key suppliers were identified, and an agreement was reached to develop new technology together. It wouldn't matter who invents. The supplier would own the material, P&G would own the field of application(s).

With some laundry chemical suppliers, labs were created on supplier sites where researchers worked side-by-side. P&G insisted that the supplier spend the same R&D % of sales investment against P&G expenditures as the supplier invested overall.

 One supplier, BASF, came forth with a novel stain removal pad, which was launched by P&G as MR. CLEAN MAGIC ERASER.

Over 2200 ideas for new products and important new uses of P&G and external technologies were generated and entered into our Innovation 2000 database. Numerous technical problems were solved right on the exhibit floor.

Based on feedback from internal and external attendees, and conversations with many of the key business managers in attendance, Innovation 2000 exceeded everyone's expectations. The goal to deliver one, actionable, significant and entirely new product idea in each Business Unit was easily surpassed.

- There were 110,000 hits on the InnovationNet website the week of Innovation 2000 - employees providing solutions to technical problems, watching the simulcast, and recording their ideas for new connections.

- Based on feedback, the external exhibitors identified dozens of new connections where their technologies could solve P&G problems and form the basis for new product benefits.

- Several Confidential Disclosure Agreements were prepared on the spot by our Technology Acquisition Group, which also facilitated several technology transfer discussions.

**One Overall Conclusion was clear.…We had established that OPEN INNOVATION was here!. P&G was wide open to external inputs, and P&G employees were charged to reach internally and externally to make technology connections.**

A majority of the ideas presented were early stage technologies that would take many years, and probably longer, to implement. Further, it is never a straight line from inspiration to execution. But, could we trace what might have happened with some of the ideas presented?

Lee Ellen Drechsler, currently head of Corporate R&D Platform Technologies, recently searched some 100 reports from technologists following INNOVATION 2000. She sought to identify specific interest at that time in particular leading edge technologies researchers encountered at the event.. Then, using subsequent 20+ years of regular "Smart Learning" reports followed what, if anything, might have happened. The findings were very impressive, and highly encouraging. To illustrate a few examples:

**SENSORS:** The Australian Membrane and Biotechnology Institute (AMBRI) presented the possibility of biosensors for detection and measurement of DNA, RNA, enzymes, and microorganisms. Such sensors are now much more accessible and used routinely across the Company during product development and clinical testing, as well as for claim support. The tremendous breakthroughs in genomics over the past two decades have enabled biosensors to go from a "What if?" idea at INNOVATION 2000 to a set of measurement tools that we rely upon.

### ENERGY AS MECHANISM TO DELIVER BENEFITS:

The Maxwell Company exhibited their Purebright sterilization technology for antibacterial benefits for tooth care. Although not a direct path, the use of LED light to activate CREST WHITE STRIPS evolved and led to market products combining the LED technology with both the CREST WHITE STRIPS product and with a WHITENING EMULSION applied product.

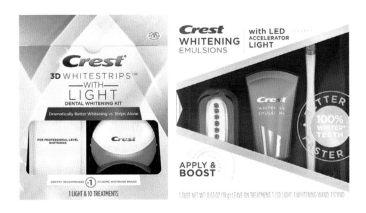

### INTERNET-ENABLED PRODUCT RESEARCH:

The Oral Care group originally developed a web-based data capture technology that enabled consumer studies to be shifted from central research sites to individual dental and physician's offices, resulting in higher quality learning and easier access to the right panelists. The IAMS people (owned by P&G at the time) then extended that to in-home pet research. Now, much of today's research is done using "connected home" technology where passive, privacy-safe sensors and mobile phone apps capture behavior and responses during product use among willing panelists."

**ENCAPSULATION.** This is a critical technology area for being able to capture and deliver the right fragrances at the right time for a given application. It is important for a wide area of laundry and household products. 3M showed core/shell microcapsules that could deliver both water soluble and oil soluble compounds. Fabric Care researchers had also presented the progress they had made encapsulating perfume within polyvinylalcohol (PVA), a water-soluble material. Studying the performance of these external and internal capsule prototypes,  including their problems, inspired further searching for the right shell material and eventually led to the melamine capsules first launched in Tide detergent.. These melamine capsules had great application to laundry washing and drying situations because they were tough enough to withstand the laundry process, were retained well on fabric during the wash cycle, and would break and release fragrance under light force once deposited on garments. Perfume microcapsules went on to be a key enabler for the successful UNSTOPPABLES laundry fragrance additive product line.

Lee Ellen reports that the "Poster Session" approach used in INNOVATION 2000 is used widely in the Company today, both within and across very diverse Communities of Practice, and almost always includes external partner inputs. It's clearly a proven way to effectively connect and transfer ideas.

But, stepping back from this, what were the main learnings from this event?

- **Despite today's communication technology, the barriers to innovation connections are truly very significant.** Both of the parties - the owner of the technology and the one in search of a solution - must be proactive and persistent in their mutual search for connections. Although the INNOVATION 2000 event was years ago, the principles behind it to emphasize the importance of connections is still extremely relevant. One can't create a major INNOVATION event on any kind of a regular basis. But whatever the mechanism might be to achieve, the importance of connections has to be internalized and reinforced within the research community.

- Non-obvious connection-making requires more energy than obvious connections. It's clearly best when a large community of participants are involved, as in INNOVATION 2000…. where possible connections were increased by the breadth and diversity of participants. But on an ongoing basis, it needs both technology seeker and provider to be proactive in reaching for opportunities to explore possibilities.

Companies will have to explore the best ways to encourage and achieve external innovation connections within their organizations. It's not easy, and unfortunately I found one route NOT to pursue. Unfortunately, it occurred within P&G.

Several years after I retired, I learned that P&G had establish an internal goal for their researchers and marketers "We will partner 50% of our innovation with outsiders". People were tasked with achieving specific external technology projects that they brought into the Company. Goals were established, and project tallies were recorded.. but, there was no business value attached to the projects, or the goal… just a number. In studying the projects themselves, it was obvious that they were small ideas…not really creating significant value. Very disappointing!,

Clearly, any goal setting on acquiring external technology, just as in developing major projects, has to be centered on big, valuable ideas.

# When small ideas and projects are rewarded, one will get lots of small ideas!

# 32.

# THE VALUE OF UNUSED PROPRIETARY TECHNOLOGY

I was very fortunate to work in a technology-rich company. In the late 1990's we owned 27,000 patents, 4,000 unique titles, and created 3,000 new patents each year. We had a world-class technical staff that included 1200 PhD's located at 19 technical centers worldwide. Needless to say…"**WE HAD THE BEEF!**"…**We had the combination of talent and the accumulation of intellectual property that was constantly being renewed.** In fact, there was more technology than we knew what to do with! We were using less than 10% of our technology base in our own products.

P&G was not unique in this regard. Our benchmarking indicated that this was fairly typical of other leading technology-rich companies and industry leaders. The reason is simple: patent law requires you to protect inventions and the company's freedom to practice. This is critical as inventions occur early in the innovation cycle, well before one knows if and how one can use the intellectual property. Waiting until you know **EXACTLY** how you'll use it is simply not an option. By then, it's too late to protect with patents. So, the IP Strategy was to discover, patent, protect and then work hard to commercialize.

Historically, P&G was VERY protective of the company's patents and know-how. They were treated as "corporate secrets" . Licensing occurred rarely. When it occurred it was viewed as the avenue of last resort. Frustratingly, we had 90% of our technologies virtually "sitting on the shelf" collecting intellectual dust!

In our overall interest in moving to an 'OPEN INNOVATION" environment -partnering with suppliers, universities, and anyone with a great idea...we decided to make a complete change in our patent portfolio strategy. Durk Jager, P&G's CEO and I announced the change in 1998. There was major apprehension among the researchers and also our business managers. But the challenge was 'USE IT OR LOSE IT". To formalize and facilitate the changes a global licensing group was set up. Their charge was to create a market across all our technology and trademark assets. If a value can be established and delivered to our shareholders, then the Company would license, sell or even give away some of its intellectual property.

 Jeff Weedman, General Manager of a newly formed Corporate Licensing Group, had his group deeply survey the P&G portfolio of patents. The goal was not only to identify patents of potential interest to external licensees, but to also identify any large collection of associated patents having major value, but needing substantial additional development.

If these patent collections weren't of interest to P&G, and not of interest in their current state to others, the idea was to donate the patents to particular universities who had the ability to further develop the technology to reach commercialization.

For P&G to commercialize the assets would require significant further technical development, moving into areas that were outside of our core competencies, and would require P&G to enter businesses that were just not a good fit.

Significant detailed studies were done to identify the most qualified university to handle the specific needs for a given patent collection. Outside experienced appraisers were able to establish a very defensible value for these "technology bundles" and P&G would receive no future revenue from the technology. It was a "win-win" for both parties

By donating these "off-strategy" patents and technologies P&G received a number of major benefits. Our inventors saw their efforts building value and making their lives' work worthwhile. We were monetizing their inventions!. Our patent maintenance costs were reduced. P&G developed stronger relationships with best-in-class research institutions. State and local governments were elated since the exploitation of the technology was expected to generate significant revenue and jobs for the university and the community. **Finally, these donations created a tax benefit for the company, and a clear return for shareholders.**

A few of the largest multi-million dollar donations were:

- 40 PLASTIC MOLDING PATENTS: MILWAUKEE SCHOOL OF ENGINEERING - These were advances in computer-aided "soft and hard" tooling, where MSOE was a leader in rapid plastic mould prototypes, with extensive pilot plant facilities. A significant school effort was put forth, and major advances were made. However.,separate industry advances in computer numerical control (CNC) eventually surpassed the technology. Overall, the experience proved very effective in challenging students and also advancing basic technical capabilities at MSOE.

- KEY PATENTS FOR "SMOOTHIES" SHELF STABILITY...KANSAS UNIVERSITY: KU had developed infrastructure for fast commercialization of food and drink products. The technology offered the opportunity for hard-to-achieve milk stabilization in packaged drinks.

- 33 CANCER PATENTS, AND 3 PATENTED CANCER DRUGS.... ARIZONA UNIVERSITY - AU had a  very competent cancer center, with particular interest in the areas we had explored, and the substance of our patents. P&G researcher Dr. James Camden had unexpectedly identified a cancer drug lead, which he creatively developed to a bank of patents having potential for treatment of HIV and Hepatitis C.

- 100 ENHANCED PAPERBOARD PATENTS: WESTERN MICHIGAN UNIVERSITY: WMU had a leading program in paper technology, with extensive pilot plant and semi-works facilities for cardboard manufacture. The technology offered an opportunity to change the game in producing stronger, lower cost, and more environmentally-friendly cardboard. This was another "out-of-the-box" idea developed by Dr. Camden, and pursued by the Corporate Packaging group through pilot plant operation.

Overall, the donation of major technology assets was one of those hard to find win-win situations for both P&G and the universities.

# 33.

# "CONNECTING TECHNOLOGY WITH STRANGE BEDFELLOWS"

As Durk Jager and I were brainstorming ideas on new innovation approaches one day, I raised the idea of undertaking joint developments with companies with expertise in strikingly different technical areas than ourselves. Trying to work with an innovator to create something that neither of us might even think about, or probably wouldn't have the expertise to create if we did.

I developed a list of possible companies, felt we needed to undertake this at the top of the partner company, and sought Durk's support to try it.

He fully bought in. Three companies at the start were Kodak (imaging), Phillips (Electro/Mechanical Devices), and Gore (Unique membranes, Goretex.). Durk formally contacted the CEO's. He and I met with them and their CTO. There was a great deal of "what's going on here" and apprehension at the start. But after discussing our thoughts, and seeing the opportunity that might be possible, their reactions turned to enthusiasm.

Appropriate disclosure and invention guidelines were signed, and cross-company teams were formed..with the charge, "Think big... forget the incremental"

It turned out that Kodak was off balance at the time with major business issues from digital photography competitIon, and the project was quickly put on hold. With Gore, a breakthrough allergy mask, and a breakthrough gum care floss were identified. With Phillips, a cellulite-reducing device with their massage system and our topical skin firming/fat burning technologies was identified.

We also opened discussions with Hasbro Toys/Creative Electronics and Scotts-Miracle Gro for gardening/pest control.

The cross-Company teams that were formed were highly energized, and appreciated the competence of their partner. Things moved quickly.

 The Gore project led to P&G's full appreciation of Gore's dental floss (GLIDE). It was superior in gliding between teeth, and freeing food debris. Further, it was highly supported by dentist professionals. Gore was the inventor and leading global expert in this waterproof membrane technology. The interaction eventually led to P&G purchasing the GLIDE product and Gore continued to manufacture the product.

The Phillips project led to a prototype machine, but the cellulite reducing goal was a stretch for the strength of the technology, and it was closed down.

With Durk's unanticipated retirement, and the changes involved in that, along with a very active new brand program needing to be launched from the Venture efforts, the "joint-effort" program was suspended.

I continue to believe there's definite value in undertaking a joint innovation program with a Company with a strong, but very different technology base. We never got the chance to fully test the possibilities, but judging the excitement and energy of the team members as they undertook the challenge in the trials we had, I think important findings can definitely be achieved.

# SECTION SEVEN

## "THE BOTTOM LINE"

### VII

# 34.

# "KEEPNG THE INNOVATION SPIRIT ALIVE!"

It's extremely important to keep the innovation spirit in a company alive - year after year- as people change, and as emphasis on different aspects of the business change. As the company grows, the innovation spirit can be very much diluted or perhaps even lost.

There are probably many ways that this spirit can be reinforced year after year. However, we had a situation at P&G that has been a major factor in achieving this, at least for the last twenty-five years.

When John Smale retired, he established the John Smale Innovation Award in 1997. He set up a trust fund with his own P&G stock designed each year to recognize current employees' technical achievements that were potential game changers.. A special achievement medal and cash award are presented to the recipients. Some of these achievements were proven, others were not, but they were all designed to have tremendous potential to change the basis for competition in important company businesses. The award also was designed to recognize employees whose work embodied "The Spirit of Innovation."

John's view of this award was captured on a video and that video has been used every year at the annual R&D global meeting when the John Small Innovation Award winners are announced.

John's message in that video drives home the spirit of innovation that he so wanted embodied in the company, and which has served to reinforce that innovation spirit ever since this award was established.

His words, with his deep and very earnest sounding voice, drives home the message.

**"It's important that the people in R&D understand the enormous importance of what they do, because it's really everything we are." "A fundamental principle I don't think is going to change and hasn't changed in our history - and that is**

## We are an R&D based company!

**We're a company whose progress and fortunes are based on the success of inventing new brands, new processes that are really distinctive, that are market changers and really revolutionize the market when we go into it.**

**"If this company is going to be successful 50 years from now, it will be successful for the same reason that it is now, and that's because we are ahead of the world in almost every category on product innovation."**

Although the use of this video was not planned at the start to provide a way to underscore the importance of product innovation continually, it has very much served that purpose.

Year after year, researchers leave the meeting having heard this message - and seen the awards recognizing potential "game changing " innovations - energized with the importance of their innovative work in driving the company forward.

# 35

# "THE SMELL OF THE PLACE"

When I started my first job in Research & Development, I commented that my job selection hinged on my reaction to "THE SMELL OF THE PLACE".

As we conclude this book, I come full-cycle trying to define the characteristics, feelings and emotions of an ideal innovating organization. What are those tangible and intangible elements that combine like a chemical reaction to make "THE SMELL OF THE PLACE" palpable.

- It should project **WINNERS.** Top quality people who welcome working with other brilliant colleagues and coming in first. Unrelenting in their drive for the perfect product, and never accepting failure, they persist until they achieve their goal. They thrive on seeking game-changing ideas. They exude confidence and mastery. They are the standard setters for their organization: capable…challenging …confident.

Ed Artzt, former CEO & Chairman of P&G, was passionate about preaching "winning". People weren't born winners, but needed to learn to be winners. A lifetime focus leading to "ACHIEVING GOALS, and BEATING EXPECTATIONS as a way of life."

- There is a a deep sense of **TRUST.** Bosses who actually felt responsible for and helping you succeed, and for shaping jobs and objectives that are exciting and stretching. People really caring how you are doing outside work. An environment where you don't need to waste time questioning motives. People who were trying to do "the right thing".

- People should proudly exhibit **OWNERSHIP.** They would be counted. It's always a fine line between success and failure - but people wouldn't hesitate to step up at critical times to take extra steps to achieve success. Company and personal success lines are blurred - they achieve them simultaneously. There is freedom to try - and there wasn't anything that couldn't be done.

- Deep **PASSION** is pervasive. A real dedication to do whatever it takes to reduce the inherent uncertainty in their project. A deep understanding that there will be a continuum of failures before reaching their project success. In my world, it could have been leaky diapers, dirty dishes, runny noses, bitter coffee, or a myriad of other everyday nuisances. But whatever it is, it takes deep passion to get to the goal.

So, how does one know that you've done a good job building an innovative organization, or are making progress in getting there …The easy test is….you'll know by an unbiased reading of "**THE SMELL OF THE PLACE**". It's actually very positive and unmistakeable!

# APPENDIX

## P&G CEO TALKS

# I.

# "IMPORTANCE OF INNOVATION"

## JOHN G. SMALE, P&G CEO (1981-1990) from "FINAL ADDRESS TO P&G MANAGEMENT (11/3/1997)

Nothing, it seems to me, has been more important to the success of Procter & Gamble over the last 40+ years than product and process innovation. Without question, the fortunes of this Company have been built on innovative products - TIDE, ARIEL, CREST, HEAD & SHOULDERS, PAMPERS, conditioning shampoos, PRINGLES, ALWAYS, CPF & CPN paper technologies, and continuing into the future, with a great pipeline of innovations.

The one thing we can be absolutely sure of tonight is that fifty years from now, when none of us will be around or certainly involved with

Procter & Gamble, if this Company has continued to grow and prosper, it will be because of product innovation. **Nothing else we do - nothing- good advertising, lower costs, better distribution - financial planning - can substitute for product innovation.**

Our products are our work. The significance of our work is our ability to invent and keep providing vital products that do a better job than anyone else's in serving consumer needs.

I take great pride, and so should you, in the fact that this Company has been the innovative leader in the vast majority of product categories in which we compete around the world. But, unless we can maintain that kind of record, the Procter & Gamble Company of the future is going to, at some point in time, lose its momentum.

So, you should assume that these enormously important assets we own- these brands - are perishable. But, even if they were not, it's entirely unlikely that you're going to double this business, even with the further geographical expansion of existing brands, without significant volume from new brands. The moral: **Innovation, expressed in new brands, as well as strengthened existing brands, was key to P&G's past. And so it will be - to its future.**

# II.

# "GAINING AN EDGE OVER COMPETITION"

**EDWIN L. ARTZT, P&G CEO -(1991-94) from "THE GLOBALIZATION YEARS" - 2023**

I don't think there is another company in the world that places so much importance on product superiority. Consistently, the Company's periods of greatest business progress are closely aligned with products that have enjoyed performance advantages over their competition.

Again and again, Procter & Gamble has set new performance standards with breakthrough technologies. Each product is not only a step change in performance and convenience, but also a major milestone in the Company's history of growth.

Innovation has been building virtually **every one** of our businesses. In each instance, we've achieved an edge over competition because our innovations have met important consumer needs that lead to preference for our brands.

It's no secret that P&G's longevity is the result of a continuing commitment to product innovation. At P&G, you just don't plan for obsolescence, you work to make it happen. And when it does, you want to be the first to market the product answer.

# III.

# "PERSONAL LEADERSHIP MAKES INNOVATION HAPPEN"

## JOHN E. PEPPER, P&G CEO (1995-98) from
## R&D GLOBAL MANAGEMENT MEETING - 12/4/2021

Personal leadership is critical to make things happen. I've seen it throughout the Company. Nowhere have I seen it more vividly than in R&D.

I've been fortunate to have worked with strong R&D leaders throughout my career. Beyond the expected qualities of any leader, there is a simpler one to describe these R&D leaders: the three C's - COMPETENCE, CHARACTER & COMMITMENT. It culminates in the drive to not allow ANYTHING to deter them from achieving the result that they believe is critical.

You uniquely have the technical & engineering COMPETENCE in providing new benefits and new innovative products

You have the CHARACTER to always do the right thing. Courage, integrity, and persistence - particularly when you're the minority view on the decision.

I was on the Disney board with Steve Jobs. I was asked what I thought was his most unique characteristic. I answered "Maniacal COMMITMENT to Excellence". I've seen it in abundance at P&G and always behind every great success we've had.

It took "maniacal COMMITMENT" to revolutionize the hair care category business with 2-in-1 shampoo, to overcome the challenge of having effective bleach performance in cool water, to enable Bounty to become one of the great brands in the world. MANIACAL COMMITMENT. That's what it takes. You know it, and it will always be that way.

# IV.

## "BREADTH OF BUSINESSES PROVIDES ADVANTAGE"

**DURK I. JAGER, P&G CEO (1999-2000) from P&G 1999 ANNUAL REPORT**

The first key to faster growth, greater business vitality, is increasing the pace of innovation at P&G. This has been true for us in the past and is just as true today.

P&G is unique when it comes to innovation. We compete in nearly 50 categories - laundry products, toothpaste, paper towels, personal cleansing, cough & cold, bone disease therapies, snacks, diapers, cosmetics, - and many others.

Some people argue that such a diversity of categories leads to a lack of focus. We see it differently. The breadth of our business enables us to connect technologies from seemingly unrelated businesses in unexpected ways.

We don't leave these connections to chance. Our Technology Council brings together R&D leaders from our existing product categories to more quickly transfer technologies from one business to another. Even as the Company grows bigger and bigger, the Technology Council accelerates the exchange of ideas much like the discussions that happened over the lunch table when we were much, much smaller.

Our Innovation Leadership Team, which I chair, is fueling our growth in new categories. It funds promising ideas that fall outside our businesses, from seed-level investment all the way through test-marketing. Previously, these kind of ideas would often go undeveloped.

Connections create breakthroughs. We are launching more new-to-the-world products than any time in our history - products like Febreze, Swiffer, our disposable mop, Downy Fabric Refresher, and Dryel, our home care product for dry-cleanables.

# ACKNOWLEDGEMENTS

We are immensely grateful to the countless individuals who helped with encouragement, project details, and also those with immense capability and determination to actually create the superior product innovations.

To Mary Ellen Brophy and Mary Jane James, our loving life partners, who provided encouragement, ideas, and also patience during the long book-creating journey.

To our children (Christine, Pamela and Meggan) along with (Kathryn, Michael, and Elizabeth) and the eleven total grand- children that make us so proud, and inspire us to create memories of that passion for work that formed out careers.

To John Pepper for his strong support to document the innovation principles and case studies so important to P&G's success, and to Jon Moeller, P&G's current CEO, for his endorsement of this perspective to drive P&G forward today.

To Victor Aguilar, the current P&G Chief R&D and Innovation Officer, who was critical and very effective. The leadership team of Marc Prichard, Damon Jones, Ken Patel and Julie Setser. The review team of Molly Marburger, Tracey Long, Erica Noble, Robyn Schroeder, Heather Valentino, and Becky Pennington. These P&G teams, along with P&G's Chief Historian & Archivist, Shane Meeker, and archivist Greg McCoy, all together provided P&G review, photos, and the interface and leadership for securing P&G agreement on use of Company copyrighted photos, package artwork, and historical photos.

To colleagues, and highly accomplished P&G R&D innovation leaders, (following time flow in book) Nabil Sakkab, Claude Mancel, Warren Haug, Kathy Fish, Ceil Kuzma and many other highly experienced and diverse global business R&D leaders for their perspective.

To Greg Icenhower, Jim Stengel, & Hank Feeley for sharing their experience with book creation and effective communication.

To Jane Walton for her expert transcribing work, and Mark Herndon of Autobahn communications for experienced manuscript compilation.

To Elizabeth Grim for her outstanding artistic design work

To early reviewers, Jack Brown, Hank Feeley, Dean Ian Robertson, Dean Marco Pagani, Charles Larson, Gina O'Connor, Paul Trokhan and many others for their perceptive observations.

To Patrick Aylward, Frank Yuchymiu, and the entire talented staff at BOOKBABY for their excellent work and support in bringing 'SUPERIOR PRODUCTS" to life.

To countless P&G researchers, including those who were global R&D leaders during the time the learnings shared in the book were created, as well as those who enthusiastically inputted information on major innovations....(following the time flow in the book)...including, but not encompassing all - Ed Artzt, Richard Andre, Gary Booth, Charles Broaddus, Charles Wosaba, James Camden, Peter Morris, Ken Morton, Gil Cloyd, J.P. Jones, Michael Jensen, John Heino, Rainer Schoene, David Digulio, James Monton, Eric Armstrong, Gordon Hassing, Irv Simon, Keith Triebwasser, Sharon Mitchell, Doug Moeser, Michael Tafuri, Robert Greene, Larry Huston, Keith Grime, Richard Stradling, Gail Moore, John Walker, Yen Hsieh, Bob Gill, Chuck Hong, Franco Spadini, Bill Connell, Paul Trokhan, Jeff Hamner, Alvaro Restrepo,

Paul Russo, Wilbur Strickland, Toan Trinh, John Walker, Craig Wynett, Ron Cramer, Patrick Masscheleyn, Sheku Kamara, Rosa Hernandez, Doug Melton, Kamilah Gillispie, and Lee Ellen Drechsler.

# V.

# PEOPLE INDEX

# VI.

## PRODUCT INDEX

# VII.

## GLOSSARY

**ACETOMINOPHEN** – An OTC drug used to treat mild to moderate pain and reduce fever.

**ACRYLIC** – A common petroleum-based thermoplastic, fabric, fiber or paint base derived from natural gas.

**AGM** – Polyacrylate -Absorbent gel material . Absorbs and retains water in diapers.

**AMYLASE** – Starch specific enzyme.

**BISPHOSPHONATE** – A class of chemicals that slow the loss of calcium in bones.

**BUILDER (Detergent)** – Mineral salts (e.g. phosphate) that neutralize water hardness (calcium, magnesium) to avoid interference with soil removal ingredients in a detergent.

**CATALYST** - Substance that increases the rate of a chemical reaction.

**CELLULASE** – Cellulose specific enzyme. Removes pills, or loose fibers on surface of worn fabrics during laundering.

**CHROMATOGRAPHY** – A laboratory device and technique that separates a compound into its chemical components.

**CYCLODEXTRIN** – Sugar molecules bound together in chemical ring structures of various sizes.

**DELTA LACTONE** – An organic compound which is a natural component of butter flavor.

**DIACETYL** – An organic compound whose volatile components are prominent in butter flavor.

**EMULSION** - A mixture of two immiscible liquids in which droplets of one are dispersed in the other, like an oil/vinegar salad dressing.

**ENCAPSULATION** – The action of enclosing one material with another resulting in a capsule.

**ENZYME** – A biological material, almost always a protein, that accelerates the rate of a chemical reaction in a variety of chemical, biological or cleaning applications.

**ESTER** – A molecule made up by bonding an alcohol with an organic acid.

**ESTERIFICATION** – The chemical process of forming an ester.

**ESTROGEN** – A steroid hormone associated with the female reproductive organs and responsible for the development of female characteristics.

**FAM** – Flexible absorbent foam made from a variety of polymers having extraordinary absorbing and liquid retention properties.

**FATTY ACID** – The components of any common food or animal fat (triglyceride) that are joined in a trio to a glycerin base.

**FERMENTATION** – A biological process in which carbohydrates are broken down into alcohol and water. It is the primary process to make bread, cheese, beer and wine.

**FUNGAL FERMENTATION** – Fungi are used to create foods and beverages in indigenous cultures by breaking down carbohydrates into kimchi, pickles. Tempeh, etc.

**IBUPROPHEN** – An OTC non-steroidal, anti-inflammatory (NSAID) drug that provides relief for pain, inflammation and fever.

**LIPASE** – An enzyme that breaks down triglycerides into free fatty acids and glycerol.

**MANGANESE** – A basic element (Atomic no. 25) that helps at trace levels to activate enzymes in the body to break down carbohydrates, fats and proteins. Also an oxygen bleach catalyst.

**MELAMINE** – A chemical made from urea. When combined with formaldehyde it forms a durable resin used in cookware, dinnerware and laminated films.

**METHYL KETONE** – A chemical material having a broad range of uses as a solvent, cleaning agent and component in lacquers and other coatings.

**NOBS** – Nonanoxybenezene Sulfate. Activates oxygen bleach to form a highly active peracid bleach.

**NONWOVEN** – A fabric-like sheet or web structure bonded together by entangled fibers or filaments by chemical, mechanical, heat or solvent.

**NTA** – Sodium Nitriloacetate; alternate detergent builder.

**OSTEOPOROSIS** – A bone disease that develops when mineral density declines and bones become more fragile.

**PAGETS DISEASE** – A disease that disrupts the body's normal process of rebuilding bone, forming abnormal bone structures, and weak and brittle bones.

**PERACID** – an effective oxygen bleach, which only forms in solution, and is not stable in the atmosphere.

**PHOSPHATE** – A form of phosphoric acid that is an essential nutrient in the body and also used historically in detergents as a builder to counter water hardness.

**PHOTOPOLYMER** – A light activated polymer that changes its physical properties when cross-linked to create a harder material.

**POLYACRYLATE** – Synthetic resin formed by the polymerization of acrylic esters.

**POLYDIMETHYLSLILOXANE** – A silicone derivative used as a hair conditioning material in shampoos.

**POLYETHYLENE** – The most widely used plastic polymer in packaging, plastic parts, and films.

**POLYVINYL CHLORIDE (PVC)** – The original plastic synthesized in 1872 used in a variety of applications, including pipe, and the 3rd most produced plastic in the world.

**PROBIOTIC** – Beneficial bacteria source in the digestive tract. Examples are yoghurt, fruits, green vegetables, chicory, oats, barley and wheat.

**PROTEASE** – Protein specific enzyme.

**PYROPHOSPHATE** – chemical additives used to treat municipal drinking water and found abundantly in the body. The original detergent builder to neutralize hard water chemicals during washing and laundering.

**RISEDRONATE** – P&G's proprietary bone loss pharmaceutical.

**ROLLER MILLING** – Cylindrical rollers to produce minor deformation in a film to create stretchability.

**SKUNK-WORKS** – Originally Martin-Marietta's secret organization (1939) that produced many war planes with engineering independence, and extreme secrecy. Today it refers to an R&D group operating separate from the main organization with great independence.

**SODIUM PERBORATE** – Oxygen bleach.

**SODIUM PERCARBONATE** – Oxygen bleach.

**SORBITOL MONOSTEARATE** – An emulsifier made from corn sugar sorbitol combined with stearic acid. It is used in many food applications to stabilize emulsions.

**SURFACTANT** – A molecule having a hydrophilic (water loving) and hydrophobic (oil loving) parts that reduce surface tension and allow the mixing of immiscible materials.

**TAD (aka CPF)** – Through-air drying; a paper making process in which excess moisture is removed by blowing warm, high-pressure air through the paper.

**TAED** – Tetraacetyl Ethylene Diamine, an activator of oxygen bleach which forms a more active peracid bleach in the washing process.

**TATAMI MAT** – Traditional floor covering and bed commonly used in Japanese rooms and homes.

**TECHNET** – P&G's proprietary Intranet that allows dialog between individuals and team members with the intent of sharing new learning.

**TOW FIBER** – A type of textile fiber characterized by its long continuous strands.

**ULTRAVIOLET CURING** – UV light treatment of a liquid polymer to form a polymeric film without the use of high temperatures or excessive chemicals.